SOTHIS PRESS

www.sothis.press
1 Waterhouse Square, London, EC1N 2ST

George Stephenson: The Locomotive and the Railway

John F. Layson

SOTHIS PRESS

ISBN 978-1-0368-0320-9

This edition is published by Sothis Press © 2023. The company is registered in England and Wales. The Company Number is 14900389.

GEORGE STEPHENSON:

THE LOCOMOTIVE AND THE RAILWAY.

BY

JOHN F. LAYSON.

LONDON:
THE TYNE PUBLISHING COMPANY, LIMITED,
14 IVY LANE, PATERNOSTER ROW,
AND NEWCASTLE-ON-TYNE

EDINBURGH	. .	J. MENZIES & CO.
GLASGOW	. . .	PORTEOUS BROTHERS.
LONDON	. . .	14 IVY LANE, PATERNOSTER ROW.
MELBOURNE	. .	GEORGE ROBERTSON.
TORONTO	. . .	JAMES CAMPBELL & SON.

THIS

Centenary Memorial

OF

GEORGE STEPHENSON

IS

RESPECTFULLY DEDICATED

TO

THE COLLIERY WORKMEN

AND TO

THE RAILWAY OPERATIVES AND SERVANTS

OF

THE UNITED KINGDOM.

1881.

Preface.

GREAT events and illustrious names have crowded hurriedly within the historical confines of the nineteenth century. The truism suggests the possibility that in a concourse of famous men the most distinguished may be hidden or disregarded through the prominence given to the later arrivals upon the scene. In these pressing times, therefore, such an event as the birth of the Railway System, and such a name as GEORGE STEPHENSON'S may be lost sight of in the crush of other remarkable circumstances, and the presence of eminent after-comers within the select arena of triumphant effort.

Art and Science, Research and Literature, Invention and Commerce, Exploration and Adventure, have combined to render the record of the achievements of the past eighty years frequently brilliant and invariably interesting. Nor is the story of the invention and improvement of machinery, subsequent to the year 1800, the least attractive to the general reader, or of questionable importance to society. Mechanical Engineering, while it does not owe its origin to the period in question, has undoubtedly been nurtured and trained within that time; and although as a science it may now be said to have attained to *robust age* and stature, the epoch of its tutelage may not be more worthily commemorated than by honouring the Centenary of the first great Railway Engineer.

Of the various mechanical contrivances now in use, none have created a more thorough revolution in the condition and habits of civilized nations than the Locomotive. The offspring of the genius of Trevithick, but soon abandoned by its erratic parent, the nursling was adopted by William Hedley of Wylam, who taught it to progress without extraneous help. George Stephenson, having seen and liked the ungainly foundling, took it to his home and heart, cared for it, fostered and improved it, and then endeavoured to send it out into the world to make its own way. But strangers for a time would have none of the Killingworth engine-wright's *protegé*. Its appearance and behaviour were not prepossessing. However, the foster-father of the Locomotive, to use his own words, "would not be put down;" and the thousands of miles of railroad in constant use in all parts of the globe testify alike to the correctness of Stephenson's judgment and the unbending energy of his disposition and character.

In the following sketch no attempt has been made either to enhance the value of his labours by ignoring the meritorious claims of his contemporaries, or to give in detail a life of "one of England's greatest working men," such as has been already written by the talented author of *Industrial Biography*, to whom all who take an interest in Mechanical Engineering and its early followers must always owe a debt of gratitude. To place a narrative of Stephenson's true connection with the Locomotive and the Railway, along with the leading facts in his eventful career, in the hands of many who may have neither the time nor opportunity for the perusal of a larger work on the subject, is the aim of this Centenary-Memorial.

CONTENTS.

CHAPTER V.

CHAPTER VI.

CHAPTER VII.

CHAPTER VIII.

CHAPTER IX.

CHAPTER X.

CHAPTER I.

THE WAGGON, THE COACH, AND THE RAILWAY.

TWO hundred and forty years ago a visit was paid to a French madhouse by an English nobleman, the Marquis of Worcester. As he passed through the wards containing the cages wherein the most unfortunate of their species were confined, like so many wild animals, the attendant who accompanied him described the various hallucinations peculiar to the inmates of that abode of the hopeless. As they approached one of the caged cells, the steps of the visitor were suddenly arrested by a pitiful cry and the terrible appearance of a man whose cadaverous and care-worn countenance peered through the massive bars which his thin, bloodless hands tremulously clutched. The ashy lips, from which the cry had proceeded, again parted, and a voice, hoarse and husky, but fierce in its earnestness, exclaimed, "I am not a madman. I am the discoverer of a power of incalculable moment to mankind."

"What has this man discovered?" inquired the Marquis.

"A mere trifle," the keeper answered derisively; "but he

wrote a book about it, nevertheless. Why, you would never guess what the discovery was—to use the steam of boiling water for the navigation of ships, the driving of carriages, and a host of other miracles equally incapable of performance."

Such was the fate of him who, in all probability, first projected the idea of steam locomotion—Solomon de Caus, a native of Normandy—and such the reception which was given to a discovery calculated to confer stupendous benefits, not only upon France, but the world. The age was not, however, propitious to scientific or mechanical research. Supineness in the Court, and superstition in the Church, together with the antagonism of officials towards anything which took the shape of innovation, conspired to hold back from society for a time the advantages which have since attended the construction of the steam engine.

Confident in the soundness of his conclusions, poor Solomon de Caus had laid a description of his plans before the King of France; but the mind of the monarch was not fitted to deal with such complicated details as were therein presented, and the readiest way to dispose of the matter was to treat the Norman genius and his discovery with contempt! The influence of a prince of the Church was next solicited to favour the designs of the obscure student of Nature and her forces; but, exasperated by repeated and urgent appeals, the Cardinal consigned his importunate suppliant to the confines of a madhouse, and Solomon de Caus and his premature project were lost to his country and mankind. A similar reward might also have been meted to the foster-father of the locomotive and the railway, had the innate determination and energy of George Stephenson been less dogged and less dauntless than the event proved these to be.

Material improvements in the ordinary conditions of life

are but seldom to be attributed to accidental circumstances. Mighty and beneficial changes in the surroundings of human existence have resulted from individual effort, and the patient, persistent pursuit of progress and proficiency. And yet, in spite of the truism, men are generally averse to encounter minor evils, even when major advantages are anticipated to follow a contemplated alteration in the established customs, methods, and manners of a people. The means which are still adopted for travelling purposes by different nations afford ample illustration of man's natural repugnance to that which promises to interfere with settled usage. Individuals had to do battle for the institution of the stage coach in the seventeenth century, as well as for the formation of public railways, and the adoption of the locomotive, in the nineteenth. Happily the champions of progress in both instances were successful.

Till towards the close of the sixteenth century English modes of transit were of a very primitive character. King Alfred's State-carriage has been described as bearing a striking resemblance to a farmer's waggon. No doubt our modern ideas of kingly dignity would receive a palpable shock were a successor of that good monarch to be found travelling now in a similar vehicle, and in the act of goading a team of oxen by the aid of a long stick barbed with iron, as depicted by the pencil of the artist and the pen of the historian. However, we have no reason to think that Alfred appeared to his dutiful subjects as otherwise than dignified and happy, when thus seen riding in his uncouth and rumbling equipage.

In course of time, what had been the exclusive property of the highest personage in the State, came to be offered for hire to all who could pay for the luxury, or were too timid to mount a horse; until, in the reign of Elizabeth, the waggon, as a royal equipage, was supplanted by the coach, which had been imported by the Queen from Holland. To ensure as much safety as might reasonably be expected from the deplorable condition of the roads, a coachman of experience was also engaged in that country. This Dutch progenitor of the British race of Jehu, and his joint importation, formed important additions to the regal consequence and grandeur of the virgin Queen. The State-carriage in which Her present Majesty occasionally proceeds to the opening of Parliament was constructed for the use of George III. upon his accession to the throne, and is a slightly modernised copy of that in which Elizabeth made journeys to various parts of her kingdom, to the consternation of horses and the bewilderment of their riders.

Stage-coaches for the conveyance of those who desired to travel expeditiously were not established in this country until the year 1657, although private carriages had been in use for sixty years previous to the introduction of coaches designed for public hire; and, for a considerable time after the latter had become a necessity to their patrons, the old lumbering waggon, with its six horses and usual pace of three miles an hour, continued to hold a high place in the estimation of the English traveller. One reason for this was to be found in the very imperfect condition of the roads. Tedious as the rate of progression may have been, travelling by the antiquated means of a stage-waggon in those days was more suited to the neglected highways and the tastes of the people, than by the lighter, and consequently less safe, rival which was struggling slowly into public favour. Some idea of the distaste with which the stage-coach was at first regarded may be conceived from the suggestion of a writer who published a pamphlet in the year 1673. He says: "The multitude of stage-coaches and caravans travelling on the roads might all, or most of them, be suppressed, especially those within forty, fifty, or sixty miles of London." Feeling that his covert advice was not likely to be adopted, however, the author slightly moderates his tone, and gravely proposes that the number of stage-coaches should not exceed the proportion of one vehicle to each county town; that the coaches should individually make one double-journey weekly, returning with the same horses with which they set out; and that the rate of travelling should be limited to thirty miles daily in summer, and twenty-five in winter. The arguments adduced in favour of this series of propositions are not more ludicrous than some of the objections which were raised against the railway proposals of George Stephenson, in the year 1825! Three reasons are given in support of the assertion that stage-coaches were most mischievous institutions. In the first place, they tended to destroy the breed of good horses, and conduced to careless horsemanship; secondly, they hindered the training and employment of watermen, from which class the country largely derived its supply of seamen; and thirdly, they exerted a prejudicial influence upon the Imperial revenue. It is pleasing to note that the much-maligned medium of transit survived not only this attack, but others also of a similar character.

Early in the reign of William III. the stage-coach had arrived at a period in its history when its promoters might treat with comparative contempt the railings of its opponents. The opinion of a traveller, given in the year 1691, may be adduced in illustration:—"Here one may be transported, *without* over-violent motion, and sheltered from the influence *of the* air, to the most noted places in England, with so much speed, that some of these coaches will reach *above fifty miles on a summer day!* That little improvement had been effected in the rate of progression during the succeeding seventy-five years, we learn from the experience of John Scott, afterwards Lord Eldon, who in the year 1766 travelled from Newcastle-upon-Tyne to London in a 'Fly;' so designated on account of its *speed.* He who became a famous Lord Chancellor spent *nearly four days on the road.* The same distance may now be covered in *six or seven hours* by rail! Some old advertisements relating to coach journeys have been preserved; the following offers an illustration of the tediousness of a trip in 1742. It is taken from *Walker's Birmingham Newspaper*, of the 12th April in that year, and the public announcement is made that "the Lichfield and Birmingham Stage Coach set out this morning (Monday) from the Rose Inn, at Holborn Bridge, London, and will be at the house of Mr. Francis Cox, the Angel and Hen and Chickens, in the high town of Birmingham, on Wednesday next, to dinner, and goes the same afternoon to Lichfield; and returns to Birmingham on Thursday morning, to breakfast, and gets to London on Saturday night; and so will continue every week regularly." A journey from London to Birmingham, which then occupied nearly three days, may now be accomplished *in three hours!*

In the old coaching days considerable interest was attached to the journey of a few hundred miles. The merits, or demerits, of rival conveyances had sometimes to be studied, and preference given to the coach that was supposed to have the smaller number of accidents recorded to its disparagement. The intending traveller had usually to adjust his earthly affairs before starting; besides making his will, if that necessary qualification to a peaceful departure from this world had not previously been attained. A painful leave-taking had not only to pass between him and his nearest and dearest ones, but a number of his acquaintances and neighbours had also to be visited and hugged, and not unfrequently public prayer offered for his *safe* return. If the traveller was in comfortable circumstances, he advertised for a companion or two to share his difficulties or dangers, and the extortionate demands of post-boys and innkeepers, besides the heavy hire-charges of a chaise. The character and position of those who offered themselves as travelling companions had to be scrutinised and determined; while the journeys so undertaken frequently began with misgivings and fears, and were brought suddenly to an issue with the fractured skulls and dislocated limbs of the travellers. But a happier era dawned when the practical application and common-sense of Telford and Macadam improved the roads, and so gave encouragement to men of capital and enterprise to construct light and commodious coaches, and to improve the breed of horses, until at length the English mail coach became famous throughout the world.

Like the stage-coach of the past, the railway of the present is of English origin and growth. It began its career in a very humble manner. Before anything akin to systematic attention had been paid to the roads of the country, or Highway Boards had been devised to that end, pack-horses were used for the conveyance of merchandise, especially when the ground to be traversed was of an undulating nature, and the distance not too great. As the population and its requirements increased, canals were formed to further the interests of trade, and connect the larger centres of industry and wealth. But while these once popular channels of transit may be said to have owed their existence to the general commercial exigencies of the nation, the birth and development of the railway is primarily to be attributed to the need of our coalowners for greater facility in the carriage of the produce of their mines to the riverside wharf. Before the advent of the infant railway, coal was transported from the colliery to the shipping staith in the most primitive fashion; horses, ponies, asses, and mules being utilised for the purpose, and the burdens being carried in sacks or panniers, which were slung across the backs of the animals. Those who have been familiar during their lives with the heavily laden coal-train, consisting of twenty or thirty waggons drawn by a powerful locomotive, cannot adequately realise the difficulties in transit which preceded the laying down of the first line of rails.

The first colliery waggon-ways consisted of rude planks of wood, which were placed upon the tracks in order that the waggons might travel with less resistance than upon the uneven ground. The next step in advance was the addition of wooden rails with a raised rim to keep the waggons on the track. Further improvements were effected by placing thin plates of iron upon the wooden rails, and making flanged wheels for the waggons; the latter rendering the raised rim or flange upon the rails unnecessary. In course of time, the idea was naturally suggested that rails wholly made of cast-iron possessed advantages over those of a composite character. Cast-iron rails, in turn, gave way before the superiority of malleable iron; and to-day steel rails are required to resist the great strain of the immense traffic which passes over certain portions of the principal lines of the country. At the beginning of the present century, Mr. Benjamin Outram introduced "props" of stone for supporting the joints and ends of rails; and his plan was generally followed in the construction of new colliery "plate-ways." The roads thus formed were known as "Outram roads;" or for brevity, "tram roads." These were the immediate predecessors of public railways; the stone "prop" being superseded by the wooden "sleeper."

Horses were generally used in the haulage of waggons upon the early "plate-ways;" but from time to time other descriptions of propelling power were proposed, tested, and found to be mainly impracticable or totally useless. Various engineers had endeavoured to carry to a successful issue James Watt's design of a locomotive for common roads. Much time, ingenuity, and money were expended over those sanguine but futile attempts to construct a reliable "travelling engine;" numerous experiments were made; many patents were granted. Repeated failures, however, caused the idea of a steam carriage, available for the common highway, to be abandoned by engineers and capitalists as utterly hopeless. Baffled in their endeavours in one direction, engineers turned their attention towards another, and, happily, with greater encouragement and better promise of ultimate success. Trevithick, a native of Cornwall, appears to have been the first to demonstrate the practicability of a union between the locomotive and the railway. Unfortunately for himself, he did not possess the natural qualities necessary to bring the idea he had conceived to perfection. He was erratic and restless, and lacked the spirit of patient, persevering industry invariably to be found in a successful inventor. It is not surprising, therefore, to find the ingenious originator of the railway locomotive soon taking up other projects, and leaving other men to bring into practical use that which was calculated to enrich his position in life and immortalise his name. In the year 1804, Trevithick constructed a "travelling engine," which was placed on the colliery railway at Merthyr Tydvil, in South Wales. After undergoing repeated experiments and alterations, it was found to be defective in principle, and was in consequence abandoned. An insurmountable difficulty appeared to stand in the way of steam locomotion being applied to the railway. Mechanical engineering being yet in its infancy, it was erroneously supposed that the wheels of a locomotive would turn round without *biting* the rails; or, in other words, that its adhesive power would be insufficient to ensure the propulsion of the machine and its load along the line. Stephenson subsequently proved the fallacy of the fear; but the difficulty, though imaginary, proved a source of much trouble to the early mechanical engineers.

The first locomotive used regularly on a private railway was constructed from the design of Mr. John Blenkinsop, who, in the year 1811, took out a patent for a racked rail, into which the toothed wheel of his locomotive worked when travelling. This engine differed from Trevithick's in having two cylinders—an improvement devised by Mr. Matthew Murray, a mechanical engineer of considerable ability—and, on the 12th August 1812, it began to run on the line which had been laid down between the Middleton Collieries and Leeds; a distance of three-and-a-half miles. About a year afterwards, one of Blenkinsop's locomotives was brought to Coxlodge Colliery, in the neighbourhood of Newcastle-upon-Tyne, and its inauguration there is thus chronicled in the *Local Records* of John Sykes, under date 2d September 1813:—"An ingenious and highly interesting experiment was performed, in the presence of a vast concourse of spectators, on the railway leading from the collieries of Kenton and Coxlodge, near Newcastle, by the application of a steam engine, constructed by Messrs. Fenton, Murray, and Wood, of Leeds, under the direction of Mr. John Blenkinsop, the patentee, for the purpose of drawing the coal waggons. About one o'clock the new invention was set agoing, having attached to it sixteen chaldron waggons loaded with coals, each waggon, with its contents, weighing four tons or thereabouts; making altogether an aggregate weight little short of seventy tons. Upon a perfectly level road, the machine so charged, it was computed, would travel at the rate of three-and-a-half miles per hour; but in the present instance its speed was short of that, owing, no doubt, to some partial ascents in the railway. Under all the circumstances, it was very highly approved of, and its complete success anticipated."

The toothed rail and its companion-wheel, which were the leading novelties in Blenkinsop's locomotive, never attained such a measure of success as the chronicler prognosticated or the inventor desired The engine-gear frequently got out of order, and the rail-teeth were rendered defective, through breakage, or the presence of stones upon the line. The partial success of the machine, however, stimulated other engineers in their endeavours to effect further improvements; and, while the history of those efforts undoubtedly furnishes a formidable catalogue of failure, misfortune, and disaster, many germs of excellence were discovered through the thought and labour of several earnest plodders in the field of mechanical experiment and research, which only required the fostering care of a master-genius to fertilise, by removing what was worthless, and preserving all that was advantageous and sound.

From what has been stated regarding the advent of the locomotive, it may readily be conceived that the popular verdict was decidedly adverse to its prolonged existence. It was associated in the public mind with all that was dubious and diabolic, with everything dirty and dangerous. It was clumsy and forbidding in appearance, and, by the steam being allowed to escape at high-pressure, it screamed in a manner the most repulsive and horrible. It impregnated the surrounding atmosphere with sulphurous fumes, and blighted the neighbouring fields with red-hot cinders. It alarmed men and frightened horses with its ear-piercing noises and unwieldy gait. It belched, and hauled, and strained, and jolted, at the rate of two-and-a-half miles per hour. True, it sometimes succeeded in dragging a ponderous weight, but a team of horses had frequently to attend its movements, in anticipation of a break-down, and in order to draw it ingloriously home. Such was the nature of the infant locomotive; and it is not to be wondered at that the unfortunate people who lived near its daily track never heard its distant snorting without regretting that the monster had not been strangled in its birth, or that they dreaded its approach as the bearer of ruin to their property, and possible destruction to themselves. Accidents from the explosion of locomotive boilers, now of rare occurrence, were then comparatively frequent, when we take into account the small number of such engines that were constructed previous to the era of public railways. The following is extracted from Sykes' *Local Records*, under date 31st July 1815, and shows that considerable reason existed for popular prejudice against the extended use of the "travelling engine:"—"A shocking accident happened at New-*bottle* Colliery, owing to the boiler of the locomotive engine *bursting, from* being too strongly charged. It was the first *trial of the* machine, which was intended to draw twenty waggons, and a number of persons had assembled around it to witness its setting off. The brakesman was dashed to pieces, and another man cut in two, by the fragments of the boiler, and a little boy thrown to a great distance and killed. About fifty others (of whom some died) were most severely scalded and wounded." This and similar disasters tended for a time to shake the faith of even practical men in the ultimate success of the locomotive; while capitalists, who frequently furnished the means for carrying on abortive experiments, stood aghast at the outlay they had repeatedly staked, on the barest chance of receiving any adequate return.

While we may lament the fact that men have thought and worked, ruined their constitutions through excessive application, beggared their families by the total expenditure of their capital, and finally died neglected and unknown, in prosecuting a scheme of universal rather than of selfish interest, we cannot but admire the qualities of energy and determination that reaped for their possessor the harvest for which others had also laboured and toiled. While many of finer temperament and more refined exterior retired from the contest disheartened or beaten, George Stephenson struggled bravely on. He had been trained in a rough school, and had become hardened against the enervating influence of difficulties, and was naturally indifferent to the threatenings of defeat. Possibly he could not lay claim to such genius as was fitted to conceive the grand idea of a locomotive. His bitterest opponent could not but allow that he possessed the shrewdness and ability to discern and develop its capabilities, and the indomitable energy necessary to turn these to practical account. How he effected this momentous result is told in the following pages.

CHAPTER II.

CHILDHOOD AND YOUTH.

A HUNDRED years ago a Northumberland village received a mark of distinction that might well have been coveted by the proudest city in the kingdom. To the small and otherwise uninteresting village of Wylam, on the north bank of the Tyne, belongs the honour of being the birthplace of him who became the first railway engineer, and famous as the foster-father of the locomotive. In the humble home of a colliery fireman, George Stephenson was born on the 9th day of June 1781. The surroundings of his childhood were not calculated to form a favourable starting-point from which to attain distinction in the race of life. Distant about eight miles from Newcastle, the locality of his birth exhibited the usual indications of a neighbourhood whose inhabitants were engaged in the working of iron and the winning of coal. George Stephenson's earliest infancy was associated with cinder-banks, coal-heaps, and blast furnaces.

His paternal grandfather having hailed from beyond the

Tweed, George may have inherited from his Scottish lineage some portion at least of that plodding earnestness and untiring application that were such prominent traits in his disposition, and which gave such a well-defined individuality to his character. His father—old Bob, as he was usually called by his friends and neighbours—had previously lived with his newly-married wife at the neighbouring village of Walbottle. Mrs. Stephenson, whose maiden name was Mabel Carr, was the daughter of a dyer at Ovingham, another Tyneside village; and, while the good woman proved a faithful wife to her husband, and a careful mother to her family of six children, we are warranted in concluding that to her decidedly nervous temperament and somewhat weak constitution, were added a considerable amount of mental energy and will-power, which intellectual characteristics doubtlessly descended to her son George, who was her second child. Robert Stephenson and his wife Mabel were worthy representatives of the class to which they belonged, and in all their changes and struggles they appear to have borne a reputation for carefulness, industry, and honesty—all the more honourable to the frugal pair when we reflect upon the straitened circumstances of their early married life, and the moral destitution which at that period so largely prevailed in the mining population of England.

During their residence at Wylam, Robert Stephenson and his family occupied a single room on the ground floor of a house, which still stands, beside what was formerly the bridle post-road between Hexham and the Northern metropolis. The weekly earnings of the bread-winner only amounted to twelve shillings, consequently the poor fireman *and his better-half* were frequently put to considerable *straits while* endeavouring honestly to procure the nec saries of life. Education for their children was wholly out of the question. Their small income did not allow an indulgence in luxuries, and the advantages of school attendance were not then so fully apparent to the working classes as these are now, even to the inhabitants of colliery villages. Thus it happened that Robert and his anxious, thoughtful helpmeet were compelled to concentrate their attention upon absolute and present requirements, and leave such matters as the mental training and future wellbeing of their children to take care of themselves.

George Stephenson's father was naturally of a kind, attractive disposition, a circumstance which not only exerted a powerful influence upon the son's bearing and conduct through life, but brought within its sway, also, the neglected children of the village, to whom "awd Bob" was an authority on all matters pertaining to their daily play, as well as the chosen referee in their occasional disputes. Besides possessing many amiable qualities, Robert was a story-teller of no mean order, and this of itself was sufficient to draw numerous audiences around the furnace of the old pumping-engine, for the purpose of hearing the genial fireman detail the adventures of Robinson Crusoe, recite the hairbreadth escape of one of the characters in the *Arabian Nights*, or dilate upon the prowess of some border hero, immortalised in one of the ballads of Northumbria. These deliveries were given with suitable action and expression on the part of the adult, and received with much interest and favour by the childish auditory. Not unfrequently did it happen that the narrator's stock of romances became exhausted. In such emergencies he never hesitated to draw upon his own imagination for incidents which were likely to please the little folks. Like many of his class, Bob had a great partiality for birds and animals. He took a peculiar pleasure in watching the growth of nestlings until they were nearly able to fly, when he would secure one of the finer birds and bear it triumphantly away. In these bird-nesting excursions he was sometimes accompanied by one of his olive branches; and his son George, when he had attained distinction as a railway engineer, remembered the thrilling sensations he experienced when first held in his father's arms for the purpose of seeing a nest of young songsters, while the parent birds were searching for food.

When George had been advanced from babyhood by the advent of one and then another little stranger into the family circle, he had to take his duly apportioned share of duty and responsibility, which consisted in running on errands, taking his father's dinner to the furnace-house, and keeping children younger than himself from getting in the way of the coal waggons, as these were drawn by horses along the old wooden tram-road immediately in front of the family domicile. Thus passed the first eight years of George Stephenson's life, without the boy having learned the alphabet of his mother-tongue, and having received no other mental training than such as might be imparted through the medium of his father's fireside tales, the observation of the habits of his father's tame robins, or by watching the movements of the wheezy old pumping-engine at the colliery, and the jolting of the chaldron waggons on the old railway.

The coal becoming exhausted at Wylam Colliery, Robert and his family removed to Dewley Burn, where he obtained employment as a fireman. Though still but a child, George was now expected to do something towards his own livelihood, and a post was soon found for him as a cow-herd o[illegible] the farm of a widow, named Grace Ainslie, at Dewl[illegible]

George's mistress had the privilege of grazing her cows along the sides of the waggon-way, and it was young Stephenson's care to prevent the animals from coming in contact with the waggons, and from trespassing upon neighbouring and forbidden pasturage. The boy received twopence a day for his services; and as he was continually in the open fields, his employment may be considered as having been more recreative than laborious. With many spare hours on his hands, he now revelled in healthy out-door amusements; and, while not forgetting his father's favourite pastime of bird-nesting, and following the same whenever opportunity offered, he now gave evidence of that latent mechanical genius which was destined to achieve fame for himself, and to confer incalculable benefits upon future generations.

Along with a young companion, named Thirlwall, George devoted much boyish ingenuity in constructing models of water-mills, which he put to practical test in the small streams which abounded in the marshy neighbourhood of Mrs. Ainslie's farm, and in erecting miniature pumping-engines and winding-machines. The materials for these mechanical efforts on the part of the young engineers were always at hand in the clay-soil of the locality, and the reeds which grew near Dewley Burn. But a start having been made in wage-earning by George, he never allowed mere recreation to hinder his chance of advancement. To get on in the world was, even at that early age, the desire of the lad; consequently we find him soon starting to hoe turnips, and doing other work about the farm, for which he received the daily remuneration of fourpence. Agricultural labours, however, were uncongenial if not distasteful to him. His greatest ambition was to be employed at the colliery, near the old pumping-engine, and in view of the winding apparatus, with its alternately ascending and descending corves. The wish was soon gratified, as he joined his elder brother as a "picker;" the duties consisting in separating the coal, when it was brought to bank, from stones and other refuse materials. His weekly wages now amounted to three shillings, and a further step in promotion was gained by his being set to drive the gin-horse; a position that carried with it an increase of twopence per day, and which we also find him filling at Black Callerton Colliery shortly afterwards. This pit was distant from Dewley Burn about two miles, across fields, and George was obliged to start from his father's cottage early in the morning, in order to get to work betimes, and to walk home in the evening when his labours were finished, under all the disadvantages of a rough and sometimes almost impassable road. But he was now a strong, growing lad; and as he trudged along, with bare feet and legs, he was more concerned about the movements of any thrush or blackbird he might see on his way than the inconvenience of "plodging" through the pools, or treading the quagmires that lay in his track. There was not a nest in the hedge-rows between Dewley and Black Callerton into which he had not peeped. The single apartment which served all the purposes of bedroom, kitchen, and parlour to the family, was also made into an aviary, through the bird-nesting proclivities of George and his father; blackbirds being especially the favourites of the boy. Unconfined by cages, the feathered youngsters were soon taught to feed from the hand, and fly about the cottage; not confining their flights to the interior. *One blackbird, in* particular, became so tame, and was *attached to* the home where it had been reared, that some years it never failed to return after having spent the pairing-season in the woods with a mate. After flying in, out, and about the doors during the day, it usually took up a roosting attitude near the sleeping forms of its human friends at night. Besides indulging his great liking for birds, George prided himself on the possession and superiority of his stock of tame rabbits, kept in a house of which he had himself been the architect and builder.

He had from his early childhood acquired an absorbing interest in colliery engines and all their belongings. When but a very small urchin, nothing gave him greater pleasure than to watch the motion of the pumping machine at Wylam, as it wheezed, and creaked, and moaned; while he listened to the plunge of the pump far down into the bowels of the earth with feelings of wonder and curiosity. Throughout his boyhood he considered the position of an engineman to be the end and aim to which his whole energies should be devoted; and when, about the age of fourteen, he was taken on as assistant fireman to his father at Dewley, he gave expression to his unbounded delight, as he was now, he thought, on the high road to the attainment of his ambition. There was only one circumstance, at this period of his life, which caused him to feel at times otherwise than happy, and which served to curb the natural buoyancy of his disposition. It was feared that his age and stature would be barriers in the way of his retaining the situation to which he had been appointed. To prevent such a contingency, the young fireman had recourse to strategy; and whenever the owner of the colliery was on a round of inspection in the neighbourhood of the engine, George invariably kept out of sight, in order that his extreme youth might not be discovered by his employer. By this means he was enabled to retain his position, with its emolument of six shillings a week, until time had given him so much experience in the work that his boyish appearance was not likely to militate against the gratification of his wish to succeed in the world.

Hard necessity again compelled Robert Stephenson and his family to remove their quarters. The working of the Dewley Burn Pit having ceased through the failure of the coal supply, they removed to Jolly's Close, near to the village of Newburn. The two eldest sons were now both employed as assistant firemen, while the younger ones worked as "wheelers" or "pickers," and the two girls assisted their mother in the duties of the household. Although the father and his four sons were all earning money, we find the family still occupying a single room, the capacity of which, available for the ordinary purposes of living, was much circumscribed by the space occupied by the three low-poled bedsteads upon which the various members of the family slept. To the uninitiated in the usages of colliery life, these domestic arrangements may appear singular, if not positively indecent. But the Stephensons simply followed a practice which was then too common among their class, and which has, even now, not been completely abolished by improved cottage accommodation in our colliery villages. Besides, the customary day and night "shifts" frequently allow some member of a colliery workman's family to be at work while others of the same household are asleep; and thus are mitigated some of the inconveniences of over-crowding. When George had attained his sixteenth year, he was engaged as a fireman on his own account, at the Mid Mill Pit, also in the neighbourhood of Newburn, but without any increase in his weekly earnings of six shillings. Still, this was a step in a forward direction, and he felt pleasure in the consciousness of increased responsibility.

Although not of a robust constitution, out-door employment and healthy exercise had rendered young Stephenson strong and wiry. He entered into athletic competitions with a will, and his companions had to acknowledge his superiority in many feats of strength and dexterity; such as lifting heavy weights from the ground, or throwing the hammer. He was sober, steady, and industrious to a degree, and so far above many of the young men with whom he associated; but his total want of education kept him down to their level in other respects, and he might have ended his days as a very respectable colliery workman and nothing more, but for his determination to better his condition by application and perseverance in the course he had marked out for himself. He was conscious of his deficiency, arising from the lack of tuition, and he resolved to curtail his enjoyments and learn to read. He also purposed to make himself acquainted with the colliery engine in all its gearing, so that he might be able to fill the post of engineman, towards which he ardently aspired, to his own credit and the advantage of those who in the future might employ him. This latter resolve he proceeded to carry out at once, while he kept on the outlook for an opportunity to learn the rudiments of his mother tongue. All his information regarding passing events had hitherto been derived from the reading of others, and he longed to be able to be independent in this respect, and to gather knowledge for himself.

When George went to work at the Mid Mill Pit, he *formed a friendship* which gave an impetus to his opening career as a colliery workman. William Coe, a fireman like himself, soon attracted Stephenson's notice by exhibiting qualities of mind and disposition that were much in harmony with his own; and for the space of two years the youths worked at the same engine fire, united by the ties of a strong attachment to each other. At the end of that time, the pit at Mid Mill being closed, the two friends were sent to Throckley Bridge, where they worked a pumping-engine for some months. Here George received an advance in wages, which rendered him, to use his own words, "a made-man for life." The first receipt of twelve shillings, as a week's pay, was an event of no little importance to him; and he gleefully announced the circumstance to his fellow-workmen who surrounded the foreman's office upon that, to him, memorable Saturday evening. While he was employed at this colliery he still continued to reside with his parents at Jolly's Close, but the Duke of Northumberland's enterprise having proved a failure, another pit was sunk in the interests of his Grace, on land which lay half-a-mile to the west of Newburn Church, and between the waggon-way at Wylam and the Tyne. Robert Stephenson was engaged as fireman for the pumping-engine which had been erected by the Duke's engineer, Robert Hawthorn, at the new colliery; and the old man had the satisfaction of finding that his son George, though little more than seventeen years old, had been appointed engineman, or "plugman," under the chief engineer. The title of "plugman" is derived from the engineman's duty to see that the tube of the pump is kept sufficiently plugged when the water at the bottom of the shaft is at a low level, so as to ensure complete suction, by preventing any exposure of the suction-holes to the atmosphere. The son now held a position superior to that of his father; but his success neither inflated him with pride nor caused him to rest satisfied with what he had already acquired. His good fortune, and the reward which had attended his earnest endeavour to discharge his duty as a workman in a conscientious and efficient manner, but served as powerful stimulants to increased exertion and higher attainments.

The engine now became to him as a pet and plaything. It absorbed his undivided attention by day, and occupied a foremost place in his thoughts as he lay awake at night. He never tired of watching it in motion, nor inspected without admiration its various parts when at rest. That he might better study its construction, and know the purposes of its parts, he took the machine to pieces and examined the fittings separately. George Stephenson's engine at the Water Row Pit was always, while under his care, a model for cleanliness and efficiency; the engineer of the colliery being but seldom required to remedy defects. So thoroughly did the young engineman endeavour to master the difficulties which necessarily at first presented themselves to one ignorant of the simplest rules in natural or mechanical science, that in a very short time he had gained such a practical knowledge of the machine as enabled him to give advice and render assistance to men of more advanced years and longer experience in the work, but whose powers of observation and industry were inferior and less decisive than his own. Of such things as treatises on the steam engine he had sometimes heard; but had a book on the subject which occupied so much of his consideration been placed in his hands, he would have been unable to read the shortest word it contained. He had occasionally picked up stray bits of information from the reading of a friend beside th engine fire; but these in general had more reference to the devastating victories of Napoleon the First than to the bloodless triumphs of Boulton or Watt.

Finding himself, in consequence of his inability to read, at a dead-lock in his pursuit of knowledge, George determined upon attending a night-school. He was now eighteen years of age; scarcely a man, yet doing the work and earning the wages of one. Old and tall as he was, he hesitated not to confess his ignorance, and become as a child in the hands of a preceptor. He had considered the matter in its various aspects, and had concluded that the end he desired was worth all the cost and all the trouble—that the tedious hours to be spent at school and in study at home would fit him to work for better wages, besides enabling him to quench his thirst for information at the fountain-head. For a considerable time his evenings, when not at work, had been occupied in modelling in clay. Sometimes his models were of engines which he had seen; at other times he endeavoured to reproduce, by means of the plastic material, the various parts of engines which had been described to him. In this recreation he had taken considerable delight, since, when a little boy, he had constructed the miniature pumping-machines in the streamlets near Dewley Burn. Now he must, as a rule, give up this source of gratification and enjoyment, go through the drudgery of learning his letters and the multiplication table, and master the initiatory difficulties of pot-hooks and cyphering.

Three nights every week were devoted by George to school attendance, and the remainder of his spare time to study at home. Robert Cowens, a teacher in somewhat poor circumstances at Walbottle, was his first instructor; and, although neither the matter nor the manner of the lessons would be tolerated in the days of School Boards and taxation for educational purposes, it must be admitted that the homely bearing of the schoolmaster gave confidence to his pupils, and encouraged them in their monotonous studies. Stephenson's fellow-scholars were for the most part young miners, and the sons of labouring men in the neighbourhood, who desired to retrieve some of the opportunities for learning which had been lost to them in their earlier years, either through the poverty or indifference of their parents or their own neglect. Being eager for knowledge, and most diligent as a scholar, George learnt to read in a comparatively short time; and when he had attained the age of nineteen, his progress in the art of writing enabled him to trace slowly the letters of his own name—a feat in his career under poor Robert Cowens of which he was not a little proud. But he grudged the time expended in walking to and from Walbottle; and in the winter of 1799 he was enabled to prosecute his education under more favourable circumstances as to the locality of his school, and considerably better auspices as regarded the qualifications of his teacher. His second schoolmaster was a Scotchman, named Andrew Robertson, who, at the close of the last century, opened a night-school at Newburn, and only a short distance from his home at Jolly's Close. George's kindly nature at first revolted against the thought of being separated in his lessons from him who had first directed his steps in the path of knowledge. Still, the consideration that the Newburn mentor had a good reputation as an arithmetician, and the thought that under the new master he was likely to attain to a higher standard of proficiency, reconciled him to a change which was fraught with many and important advantages.

Andrew Robertson took a great interest in the advancement of his pupil, as well on account of the indomitable industry which the young engineman displayed in his studies, as for the reason that George soon gave evidence of considerable aptitude for the practice of arithmetic, a science in which the Scotch dominie himself delighted and excelled. Improving every moment of his spare time, George stood by the engine fire and worked out the sums which had been set for him by his master, to whom the well-used slate, upon which the problems and their solutions appeared, was taken each evening, and other work put down for the following day. Endued with a willing mind and untiring zeal, George now made rapid progress, and soon left behind others who had started with him from the same point in the race for knowledge. Nor is the fact of his marked success as a scholar to be wondered at, when we consider the thoroughness of his disposition or the force of his character. He could never leave any task unfinished, or rest satisfied with a partial success. To accomplish all that he knew would benefit himself in the attainment, formed an object to which he directed all the energy of his ardent and unconquerable nature. In seeking to raise himself in the social scale he conferred inestimable favours upon humanity.

CHAPTER III.

THE COLLIERY ENGINEMAN.

GEORGE now felt that he was getting on in the world; but that there was another rung on the ladder of proficiency which he must reach, he had fully and wisely determined. Friendship came to his aid and furthered his design to learn the act of brakeing an engine. The duties of a colliery brakesman are of a monotonous but important nature; those men being invariably selected for the post who have earned for themselves a reputation for punctuality, carefulness, and sobriety during their probationary training as firemen. Being one of the higher kinds of colliery labour, it is comparatively well-paid; therefore George had a double motive in seeking to acquire the practice necessary to fit him for taking the position. The envy of other workmen, it is true, conspired for a time to thwart him in his efforts after his own improvement in this particular; but he was able to overcome overt acts of opposition, and to prove his fitness for the work, through the *kindness of his* friend, William Coe, who had been appointed brakesman of a small winding-machine which had lately been erected for drawing the coal to bank at the Water Row Pit.

William occasionally allowed George to work the brake, and gave him the instruction necessary to qualify him to fill a situation similar to his own. Some of the workmen, however, having objected to this proceeding, and one having even stopped the working of the colliery on this account, the young brakesman, whose friendship for Stephenson had prompted the kindly action, took the earliest opportunity of justifying his own conduct in the view of his superiors. A banksman named Locke would appear to have taken a leading part in opposing William Coe, and in seeking to prevent him from initiating his friend into the secrets of his daily work. Coe resolutely resolved upon bringing the dispute to an issue, however, by calling upon George to take his place at the apparatus, just as the manager, Mr. Nixon, was one day approaching the brake-house. Locke, thus challenged, ceased working, and sat down, with the result that for the time the operations of the colliery ceased. Mr. Nixon having demanded an explanation of the untoward proceeding, was informed by the arbitrary banksman that Stephenson was incompetent to do the work of a brakesman, and falsely added that George was much too clumsy in his movements ever to learn. The manager, shrewdly detecting the jealousy that had prompted the objection, ordered the work to proceed, without calling in question either the wisdom of William Coe in seeking to do a good turn to his friend, or the aptitude of George Stephenson for acquiring an art which was one of the stepping-stones to his career of distinguished usefulness and honour.

The friendship which had been thus tested continued to

exist between Coe and Stephenson even after their different fortunes had separated their paths in life; and for at least three years after the incident at the Water Row Colliery the intercourse of the two "marrows" was of the most intimate and cordial nature. After working together in the neighbourhood of Newburn for about three years, they removed to Black Callerton, where George so gained the confidence of his new employers as to be appointed brakesman at the Dolly Pit. He was now only twenty years old, and the promotion he had attained is the best evidence of his good qualities and character in early manhood.

He still continued an earnest student, and had become an eager searcher after information, especially such as referred to the construction or application of steam engines. His teacher, Andrew Robertson, respected him for his sterling moral qualities, and felt proud of the advancement in learning which George had made under his care; and the poor dominie, finding his night-school likely to be deprived of its most promising scholar, and a general falling-off in the number of his pupils, also removed to Black Callerton Colliery, where he had the satisfaction of ministering further to the educational necessities of one for whom he had conceived an affection almost akin to that of a father.

Up to this period George had always managed to devote some portion of his time to caring for the wants of his feathered friends. He had inherited his father's love for birds and animals; and, like his father, he could command the presence of a number of robins that flew about him at the engine fire, attracted by the bread-crumbs which he daily saved expressly for them. He had found, too, a faithful dumb companion in a dog, whose sagacity was of a high order, and frequently turned to account in bringing his master's dinner. When so engaged, the animal heeded not the snarling of any curs he might meet upon the road, and but rarely was called to defend the possession of the tin can which contained his master's meal. When the pugnaciousness of a canine passer-by was not to be avoided, however, he could give a good account of himself, and the intruder invariably found that he had the worst of it when it came to a matter of determined hostility. Upon the occasion of one of these encounters, George's messenger was only able to retain the keeping of the tin can: the dinner had been spilt; but the master readily overlooked the loss of a meal, and felt rather proud of the prowess of his dog when he was made aware of the circumstances. But the brakesman's affections were now to be devoted to a worthier object, and his solicitude directed into another channel.

Early marriages formed then, as now, the rule rather than the exception, among the hardy population of Northumbrian colliery villages; and George sought to give practical effect in his nuptials to the custom. He took up his residence in the household of a small farmer at Black Callerton, whose servant soon found favour in the sight of the young, active, and prepossessing lodger, if she had not effected that result previously to her lover being admitted to her daily society at her master's fireside. Fanny Henderson's personal attractions were enough of themselves to draw attention, and to kindle the tender passion in any young man of twenty and her station in life; and when we are informed that to her striking comeliness of form and feature were added the sweetest of tempers, a modest demeanour, kind disposition, and considerable good sense, we cannot wonder *that the* inherent shrewdness of George Stephenson *had discovered* in the "farmer's lass" qualities that were congenial to his nature, and a heart that was likely to beat in sympathy with his own.

The carefulness and forethought which guided his actions throughout his career were apparent during the period of his courtship. At a time when young men are generally influenced more by impulse than by reason, George takes thought for the comfort of his future wife and the furnishing of their home in a respectable fashion. His wages now averaged nineteen shillings weekly, but he would add to his income by devoting his spare hours to remunerative labour, and so save money for the comfortable replenishment of a cottage ere he would take the girl he so truly loved to the altar. Besides, his duties as a brakesman allowed him a considerable amount of time, which he might turn to his own account without interfering with the services demanded by his situation, and these odd times he would utilise rather than pass in listless inaction. He had learned to mend the miner's shoes; he would now try his hand at shoemaking, as well as continue to do the work of a cobbler. The first guinea saved by him from his earnings in this way was a source of great pleasure to him, as well as a stimulus to continue in his course of self-denial and industry. The story has been told how upon a certain occasion he soled the shoes of his sweetheart, and was so elated at the thought of what a capital job he had made of them, that he carried the proofs of his ability as a cobbler, during a Sunday afternoon, after the manner of a lover of the olden time, who was only too happy if fortune threw in his way the glove or handkerchief of his mistress, to keep as a guerdon of his affection and fidelity.

The attachment which had been formed between him and the amiable Fanny Henderson no doubt acted as a safeguard to George against the temptations incident to his station as a colliery workman, and assisted in strengthening his resolution to pursue an honourable and steady course in life. He has been described by his most intimate friend and daily associate at this period as being "a standing example of manly character." At the usual fortnightly saturnalia on pay-day, while his fellow-workmen were, for the most part, drinking at the village inn or engaged in the brutalizing sport of dog-fighting or cock-fighting, George was in the habit of taking his engine to pieces, examining and cleaning the separate parts, and putting all together again, ready for the work of another fortnight, after its pains-taking attendant had received further insight into its mechanism and usefulness.

Some of the young and sport-loving miners might consider the habits of the steady-going brakesman as being too strait-laced and singular to their taste, and as an evidence of a cowardly and narrow disposition; but George could give what appeared to the average minds of those who were of such an opinion, if occasion required, indubitable proof of the bravery and manliness of his character. A pitman, named Nelson, at Black Callerton, was the terror of the respectable and peace-loving portion of the community. He had earned for himself some notoriety as a pugilist, and to quarrel with him was tantamount to giving him a challenge to fight. George, in the course of his work as a brakesman, having to draw the miners out of the pit at the close of each working "shift," gave some offence to Nelson by the manner of his brakeing the machine one day, while the pugnacious bully was ascending the shaft. With a volley of oaths, as he left the cage at the top, Nelson greeted Stephenson, and made some coarse allusions to what he considered his clumsiness. Defending himself from the charge, George appealed to the other workmen to decide whether it was well-founded or otherwise; but his angry opponent would not have the matter decided by an appeal to any such form of arbitration, and threatened to assault the brakesman. George defied him, and the matter was settled for the time-being by the giving of a challenge by the miner, which was immediately accepted, and a day fixed for the fight. As the news of the projected battle spread through the village, much excitement prevailed. In regard to the ultimate result, there was but one opinion—George would be killed. Although popular sympathy was in his favour, no one could believe that the untrained and studious brakesman was a match for the man who had defeated so many in hard-fought encounters. But the event falsified the general anticipation and fears, and taught the roisterer a salutary lesson. Nelson gave up working in order to keep himself strong and able to demolish his antagonist. Stephenson, however, went about his daily employment as if nothing unusual was to occur; and, when the appointed evening arrived, he went into the Dolly Pit Field to meet his exultant enemy with all the coolness of a professional pugilist. Stripping off his upper garments, he took up his position for the first round; and, as much by the force of his determination as by the strength of his muscles and the celerity of his movements, severely punished his adversary. The young and inexperienced combatant was soon declared victorious, to the pleasure and satisfaction of the majority of the onlookers. This was his first and last pugilistic battle.

While engaged at Black Callerton he first essayed to be an inventor by turning his attention to what he considered would be an improvement in engine-brakeing. The idea which he sought to embody in a practical form was that of a machine which would be self-reversing in its action; but his models, upon which he had spent some time and patience, were eventually discarded as useless, and he reverted in his spare hours to reading and writing, so that he might become more proficient in these arts. The application which he had devoted to his abortive conception was not, however, wholly devoid of benefit to himself, for the mental training which accompanied the concentration of his thoughts upon the object to be attained contributed to fit him for grappling with more important problems, in the solution of which he was successful. When he had been nearly two years at Black Callerton, the offer was made to him to take charge of the Ballast Hill engine at Willington Quay; a post which carried with it an increased income, and enabled him to become a householder on his own account.

The Ballast Hill at Willington received its name from the material of which it was formed. Before the days of iron-built steam colliers, the craft engaged in carrying coal from the Tyne to London and other ports made their runs to the coaly river in ballast, after discharging their cargoes. Upon arriving at Willington, from which place large quantities of the Northumberland coal was then shipped, the ballast was taken from the vessels and emptied upon a piece of vacant ground in the vicinity of the Quay. In course of time a huge mound of sand and rubbish was formed, requiring engine-power to raise to the summit the waggons laden with the ballast from the ships. At the foot of this "hill" stood the fixed engine of which Stephenson was appointed the brakesman. By industry and thrift he had saved, what was, to a young man of his calling, a considerable sum of money; and Fanny Henderson agreed to become his wife whenever their future home was ready for her reception.

The ground upon which George's cottage at Willington stood now forms the site of the imposing Stephenson Memorial Schools; and to a humble but withal smartly-furnished room in the upper portion of the two-storied building the newly-wedded pair proceeded, after the marriage ceremony in Newburn Church and a visit had been paid to the infirm father of the bridegroom at Jolly's Close. George and Fanny were married on the 28th day of November 1802; and their mode of conveyance from the parental roof-tree to the bridal home was one commonly adopted on such occasions before the construction of public railways. A neighbouring farmer provided two horses, with saddles and pillions—one animal for the joint accommodation of George and his bride, the other for that of the groomsman and bridesmaid. Thus, for a distance of about fifteen miles, the wedding party travelled, until the cottage at Willington was reached, and the happy couple and their two friends began the work of house-warming.

Comfortably settled with his young wife in his new sphere, George applied himself with unabated energy to self-improvement. He was satisfied that, to ascend still higher upon the social ladder, he must by patient application qualify himself to hold the more exalted position to which he ardently aspired. He resolved to be something more than a manual worker. By carefully husbanding his spare hours he had succeeded, so far, in educating and raising himself to a point considerably above what he could have hoped for as an illiterate youth of eighteen. He must strive after greater attainments, and deserve further success. That he might understand the laws which ruled the movements of his engine he diligently studied the principles of mechanics; while he ever and anon tested, by the aid of models fashioned by his own hands, the soundness of the theories about which he had been reading. He delighted to take up uncommon speculations, in order that he might sift some grains of truth, at least, from their bulk of chaff. Having been informed that several mechanists had tried to put into practical shape the principle of perpetual motion, and had failed in their attempts, George set to work and constructed the model of a machine, the driving-wheel of which was to rotate by the action of quicksilver. The inventor certainly gave self-action to his model, but the "motion" not being "perpetual," he wisely gave up the idea for more profitable pursuits.

Ever fond of studying the larger machinery, with its cog-wheels and cranks, his attention was arrested by a household incident to a more delicate form of mechanism than had hitherto occupied his thoughts or ingenuity. While he was at work one day the chimney of his room took fire, with the result that the apartment became filled with soot and steam, and soaked with the water which had been used by kind but unskilful neighbours in extinguishing the flames. Among the mishaps which followed the disaster was to be reckoned the spoiling of a favourite clock, through the action of the steam and soot upon its wheels. George was speedily employed upon his much-prized timepiece; and his maiden attempt at clock-cleaning proved so successful as to qualify him, by popular consent, to fill the position of clock-doctor *par excellence* of the locality.

During his stay at Willington Quay he made the acquaintance of one who also became distinguished as a mechanical engineer. William Fairbairn, afterwards President of the British Association, was then working at Percy Main Colliery as an apprentice-engineman. The intercourse of the two workmen ripened into a most close and intimate friendship, and many years after, when each had earned for himself an honoured name and high position, did they remember the pleasant evenings which they had spent in each other's society at Willington. Fairbairn frequently visited at the cottage of his friend, and has spoken of the comfort and tidiness which prevailed therein, under the superintending care of the young housewife, as well as of the industry and heartiness of her husband. Upon many occasions, in order to allow Stephenson to add a little to his income, William took charge of the engine at the Ballast Hill, while the brakesman was assisting in clearing a ship of its ballast at the Quay. George also, at this period, filled up a portion of his bye-hours in last-making, cobbling, and shoemaking labours. At home or abroad he was never idle.

On the 16th day of October 1803, Mrs. Stephenson gave birth to a son, whose name in after years was scarcely less eminent than that of his father. Robert Stephenson, called after his grandfather, brought increased happiness to his parents, and his birth cemented the ties of affection which bound them to each other. As a child, Robert was his father's pride and solace; as a youth, his companion and fellow-student; as a man, his adviser and friend. The power of paternal love and the worth of filial reverence and duty were amply illustrated in the joint-career of the famous father and son.

CHAPTER IV.

THE KILLINGWORTH ENGINE-WRIGHT.

GEORGE STEPHENSON tended the engine at the Willington Ballast Hill for about two years and a half, when he left to fill a similar situation at West Moor Colliery, Killingworth, about six miles to the north of Newcastle-upon-Tyne. It was not without some reluctance that he consented to take this step, for he had been able to add to his pay while at Willington Quay by working at ballast-heaving, as we have seen. However, the consideration that he would again be engaged about a coal mine, in labour to which he had been trained in his youth, as well as the thought that he might still employ his spare time in his new neighbourhood to pecuniary advantage, caused him to decide upon making the change. Accordingly he removed to West Moor early in the year 1805, and entered upon his duties as a colliery brakesman.

Twelve months had scarcely been passed by the devoted couple in their new sphere when death severed their marriage-tie, and brought to the husband the greatest sorro of his life. The loving wife and gentle mother was suddenly taken from her partner's society and the prattle of her little son, after giving birth to a daughter. The baby lived but a few months, and George felt as if the light of his hearth and home was thus extinguished for ever. But his was not a disposition to be entirely crushed by bereavement. He had loved his spouse with all the ardour of his earnest, manly nature; and, when the poignancy of his grief had been a little dulled by time and duty, the affection which he had lavished upon poor Fanny Henderson, as a maiden and as a wife, was transferred to her surviving child.

Ere the bitterness of his loss had been assuaged, George was asked to take charge of the engine of a spinning mill, near Montrose, in the north-east of Scotland. Thinking that the change might be useful in giving him an insight into the working of machinery with which he was not yet familiar, as well as being otherwise beneficial by removing him for a time from the scene of his sorrow, he accepted the offer of the gentlemen who were interested in his appointment. After making arrangements for the care and welfare of little Robert, who was to be left at Killingworth, he started out on foot for Montrose, a distance exceeding two hundred miles. During his stay in Scotland he gave some evidence of his ready ingenuity in grappling with practical difficulties. The pumps for supplying water to the works becoming clogged, from the sandy nature of the ground, the engineman adopted an expedient which entirely remedied the evil; while the result showed that he had taken a correct estimate of the connection between a certain cause and its effect in the problem of natural science which he had essayed to solve. By causing the lower end of the pump to work inside a wooden box, twelve feet high, into which the water flowed from the upper portion of the well, the fluid could be drawn free from sand, and the difficulty was thus overcome.

After a year's sojourn at Montrose, George returned to Killingworth, travelling a-foot as on the occasion of his going northwards. Upon reaching West Moor he found that his parents had been reduced to great poverty and distress, through an accident that had befallen his father while making some repairs on an engine. The face of the old man had been severely scorched and his eyesight hopelessly destroyed. His other sons were almost as poor as himself, and consequently little able to mitigate his affliction. George had, however, saved twenty-eight pounds during his absence, and with a portion of this sum he paid his father's debts and removed his parents from Jolly's Close to his own neighbourhood, where he cared for them in their declining years.

About the year 1808, his prospects became involved in considerable doubt and uncertainty. His own bereavement and his father's misfortune had rendered him distrustful of the future; while the social state of England, thrown off its balance by the immediate consequences of a costly war, and exhibiting tokens of restless insecurity, gave but little encouragement to a workman of his genius and industry to settle in the land of his birth. The compulsory drafting of men into the navy or the militia irritated the industrious classes and unsettled trade. With many of his order, Stephenson looked with a longing eye towards America, as furnishing greater scope and security to the intelligent workman and his savings. But to an untoward circumstance at this time England owes the advantage of retaining the services of one of her distinguished sons unt a return to the undisturbed prosecution of peaceful pursuits gave an impetus to enterprise, and allowed the merits of the great railway engineer to be recognised and appreciated in his own country.

He had been reinstated as brakesman at Killingworth only a short time when the chances of the ballot demanded that he must serve his country as a militiaman. Fate was inexorable; he must either enter the ranks or regain his freedom by purchasing the services of a substitute. He adopted the latter alternative. Thus was he, at a stroke, deprived of all that remained to him of his hard-won earnings, besides being compelled to borrow six pounds to make up the required smart-money. George felt this blow acutely, but the duties of his avocation, and the scientific studies that usually formed his recreation, kept him from brooding over his loss, and he earnestly set to work to retrieve his position, as well as effect economical improvements in that department of the colliery with which he was more immediately concerned. The habit which he had acquired at an earlier period, of taking his engine to pieces on Saturdays, he still followed with advantage to himself and his employers. The practical knowledge thus gained, while immediately useful in fitting him to remedy trivial defects, gave him courage to attempt important alterations which paved the way for greater successes and weightier achievements.

The sinking of the Killingworth High Pit having been commenced in the year 1810, a Newcomen engine was erected for pumping water out of the shaft. This machine proving a failure, the advice and assistance of engine-wrights in the locality, as well as of some from a distance, were sought to render it effective in performing the required service. George had studiously watched its erection, and hesitated not to express his opinion that the engine would prove defective in its action; but the brakesman's views on the subject met with very scant attention from those who heard him; and the sinking operations made but little progress on account of the water accumulating faster than it could be removed by pumping. The workmen engaged in deepening the shaft were being continually "drowned out." So confident was George in his opinion of the cause of the defect, and his power to remove it, that he said to one of the sinkers, who had asked one day what he thought of the engine:—"I could alter her, man, and make her draw. In a week's time I could send you to the bottom." This having been reported to the viewer of the colliery, Mr. Ralph Dodds, George was speedily allowed to try his hand at that which baffled the skill of so many of much greater experience than himself. Mr. Dodds promised that he would "make a man" of Stephenson, if successful; so that the poor brakesman had now an incentive for exertion as well in the prospect thus held out as a reward of his energy, as in the desire to prove that all the self-denying industry of his early manhood had not been expended in vain.

A considerable amount of jealousy was shown by the regular engine-wrights when they discovered that the work of altering their engine had been placed in Stephenson's hands, and they did not scruple to exhibit the rancorous feelings which another's interference in their own province had evoked. But having stipulated for his personal choice of the men who were to assist him, he cared little for any ebullition of spiteful ill-will or rivalry, and threw all his energies into the perfect accomplishment of his self-imposed task. He caused the engine to be taken to pieces, and not only devised alterations in the construction, by which its better efficiency was secured, but also effected improvements in the boiler, which enabled the latter to be worked at a pressure double to that which its builders had intended should be the limit. The work of re-fitting and erecting the engine and boiler was completed in three days; and, although it was done in a rough and, what would be considered now-a-days, an unfinished manner, showed clearly that the new engine-doctor had been guided by right principles of mechanical science, both in the conception and execution of the operations.

A crowd had assembled round the engine when the hour had arrived for testing the efficacy of Stephenson's first practical essay in mechanical engineering. Some had been led thither by mere curiosity; others had gone expecting to triumph over the discomfiture of the precocious brakesman. The officials of the colliery, who had a personal interest in the success of the alterations, scarcely dared to hope that George would prove victorious. Newcomen, the engineer from whose plans the engine had been constructed, as well as Smeaton, who had made and erected it, had declared against the utility of the projected modifications. The engine, upon being set off, confirmed the majority of those who stood near in their opinion that Stephenson was a mere upstart and charlatan who had taken upon himself to meddle in a matter about which he knew absolutely nothing. As the unwieldy machine threatened by its erratic movements to bring down the engine-house, Mr. Dodds declared that it was really in a worse condition than before; but in an hour or two, when it had got into working trim, and the lessening of the water in the shaft had shown that it was now doing good service, even those who had come to sneer at the hardihood of attempting what appeared impossible, could not but admit that the alterations had proved successful in the extreme. Mr. Dodds was delighted at the result; and not only offered immediate tokens of his appreciation of the skill and ingenuity of his engineman, in a present of ten pounds, and advancement to the charge of the engine at the High Pit, but he also gave to Stephenson very encouraging promises of future advantage.

The success thus achieved proved the turning point in his career. The advice of the Killingworth engine-doctor was sought far and near regarding wheezy, worn-out machines and their renovation, until he came to be regarded somewhat in the light of a successful practitioner. It is true that the regular faculty of engine-curers looked upon him as a quack, who was not entitled to formal recognition at their hands; still, his invariable skilfulness increased his reputation and added to the number of his "patients." Nor were his services confined to the patching up of the infirm. His favourable opinion of any new pumping apparatus was an assurance of its utility and value.

In the year 1812 he was appointed engine-wright of the Killingworth Pits. His predecessor having then been killed by an accident, Mr. Dodds recommended him to the notice of the owners, who leased several collieries, and who were popularly known as the Grand Allies. The recommendation being accompanied by a flattering account of his remarkable ingenuity and industry, effect was given to the representations of the head viewer, and George entered upon his greater responsibilities at a salary of two pounds a week. The confidence thus reposed by his superiors caused him to relinquish any desire for seeking his prosperity in a foreign land.

CHAPTER V.

GEORGE STEPHENSON'S FIRST LOCOMOTIVE.

HE entered upon his new duties as colliery engine-wright with a firm resolve to adhere to that course of sobriety and conscientious painstaking which had marked his service in the lower grades through which he had passed as a workman. He also determined to effect such improvements in the machinery as were calculated to make the most of his steam power, and thereby reduce the working expenditure. That his efforts were appreciated is seen in the fact that, soon after his advancement to the post of engine-wright at Killingworth, he was still further promoted by being appointed sole engineer to the various collieries leased by the Grand Allies. These included, in Northumberland, the pits at Long Benton and Killingworth, and in Durham, those at Mount Moor, Darwent Moor, and South Moor. Under his supervision the most extensive system of engine-planes then known to be in use in coal mining was successfully carried out: the coal being conveyed by sloping planes from a distance of about two miles underground, and the furthermost engine being fifteen hundred yards from the shaft. The smoke from the boilers of the engines used in working this underground traffic was brought the latter distance by means of flues. It has also been said that he was the first to introduce tram-roads in the conveyance of coal from distant parts of the workings to the main roads in the Killingworth Pit, with small trams on which the corves or baskets containing the coal were placed. These light carriages could be conveniently pushed along the rails by boys, and the ponies and sledges previously used for that purpose were dispensed with, at a considerable saving to the owners.

Allowing no opportunity for his own improvement to pass unheeded, he increased his capability for overcoming difficulties, and strengthened his self-reliance. His thirst for knowledge still continued insatiable. Although his progress as a scholar under Andrew Robertson had been altogether satisfactory so far as it went, George resolved to prosecute his personal education still further. This he was enabled to effect through his own industry and the friendly help of a neighbouring farmer's son. John Wigham lived at Benton Glebe Farm, of which his father was the occupier. The cottage of the Wigham family being but a short distance from West Moor, Stephenson frequently took his slate to John, as he had done formerly to the Newburn schoolmaster, to have his sums checked or set down for working out in the engine-house. When the slate was not at hand, and a problem in arithmetic mentally arose before him, a piece of chalk and the side of a coal waggon answered well his purpose. The engine-wright was seldom at a loss for an expedient; while his persistency in learning, and his careful study of all the details of the question which he had for the moment in hand, met with an adequate reward in the enhanced proficiency of the eager student. His friend was useful to him in various other respects. John was not only a good arithmetician and penman, but an acute thinker, and could reason with perspicuity on matters pertaining to natural philosophy. The contact with such a mind proved of infinite value to George in his after-career; and when he had attained the pinnacle of his fame as a railway engineer, he was not too proud to acknowledge with gratitude his indebtedness to the friendship and instruction of the Benton farmer's son.

Before he became acquainted with John Wigham, George had sometimes amused himself at his leisure moments by drawing plans of engines and tram-roads, but these were naturally of a rough and sketchy character. He now gave considerable attention to that important branch of an engineer's requirements. Wigham was a very fair draughtsman; but in learning to draw plans and sections under his eye, George proved himself such an apt pupil that he soon excelled his teacher in the art. When he had attained his thirtieth year we find him just as eager for knowledge as when, at the age of eighteen, he sat beside Robin Cowens and learnt the alphabet.

So great was his industry, that in spite of the claims of his daily duties, and the time he devoted to self-improvement, he continued his old occupation of shoemaking and mending; while he had added to this the cleaning of watches and clocks, the making of shoe-lasts for the shoemakers of the district, and the cutting out of pitmen's clothes, preparatory to the work of sewing the same by their wives. "Geordy Steevie's cut" in pit suits prevailed in the neighbourhood of Killingworth long after the ingenious "cutter" had quitted this mortal scene. By these pursuits in after hours he earned sufficient money to keep his aged parents in comfort, at a period, too, when the cost of living was exceedingly high.

But George had another incentive for adding by extra labours to the income he derived as the colliery engine-wright. His son Robert was now growing a fine and intelligent lad, and having proved in his own case, by painful experience, the depressing influence of the want of education upon a poor man's prospects in life, he earnestly determined that no self-sacrifice would be too great, could he but effect thereby the intellectual elevation of his boy. Robert had been sent to a school at Long Benton, where he was grounded in the rudiments of education; but the father knew, by having personally superintended the lad's studies at home, that the tuition imparted by the country schoolmaster was of the most meagre description, and he resolved that Robert should receive all the benefits derivable from the instruction of one of the best masters in Newcastle-upon-Tyne. At a meeting held in celebration of the opening of the Newcastle and Darlington Railway, on the 18th of June 1844, George Stephenson alluded to this period of his life, and his self-denying efforts for his son's welfare, in the following words:—"When Robert was a little boy I saw how deficient I was in education, and I made up my mind that he should not labour under the same defect, but that I would put him to a good school and give him a liberal training. I was, however, a poor man; and how do you think I managed? I betook myself to mending my neighbours' clocks and watches at nights, after my daily labour was done, and thus I procured the means of educating my son." The paternal solicitude and forethought thus manifested were conspicuously rewarded in the brilliant career of Robert Stephenson, who, when a boy, rode daily upon a donkey from Killingworth to Newcastle, where he attended Dr. Bruce's school in Percy Street during the day; returning in the same manner in the evening to his father, who watched, as the weeks and months flew past, with glowing pride the boy's progress in learning, as well as other proofs which his son gave of increasing intellectual strength and power of observation.

The interior of Stephenson's cottage at Killingworth resembled more the shop of a dealer in curiosities than the home of a superior colliery workman. The walls were covered with models of self-acting planes, engines, and machines of various kinds, from those of the most useful and practical character to some that could only be considered as utopian and valueless. The latter were all that resulted from his fruitless attempts to solve the problem of perpetual motion; the former he still hoped to utilise as opportunities arose for their being embodied in future improvements. Thus surrounded, with his son by his side learning the tasks for the following day, and a favourite blackbird looking inquisitively upon the scene, George continued to mend and clean, and cobble, and study after his regular labours were over. In this way did he serve an apprenticeship which fitted him to become a mechanical engineer of eminence and fame.

Seventy years ago, the revival of the idea of applying steam locomotion to the conveyance of coal over the tram-roads of the country arrested the attention of all who were pecuniarily interested in a successful solution of the question At the beginning of the century Trevithick had constructed a high-pressure locomotive, which was considerably superior to all its predecessors; and although that engine had been abandoned on account of its defects, still enough had been accomplished by its clever but erratic inventor to pave the way for others who might have the desire and energy sufficient to bring into practical shape the crude ideas of the impatient Cornishman. Mr. William Hedley, at Wylam Colliery, where George Stephenson's early years were spent, was the first to put Trevithick's plan for a locomotive engine to a thorough test, and to prove the fallacy of such a machine requiring "friction-wheels" to further its adhesion to the rails; which notion Trevithick acted upon in common with his contemporaries. But this was accomplished by the Tyneside inventor only after attempts and failures sufficient to have deterred one less sanguine than himself as to the utility of steam locomotion being ultimately established. In October 1804, John Whinfield, of Pipewellgate, Gateshead, constructed for the colliery a locomotive, from plans furnished by Trevithick. That machine never left Whinfield's foundry, as it was considered too light for the work it would be required to do; and it was not until Blenkinsop had introduced the racked rail in 1811, that Mr. Blackett the owner permitted a renewed attempt to introduce the locomotive upon the Wylam waggon-way. The second engine proved likewise a failure; for after it had been removed from Gateshead, when finished by Thomas Waters of that town, and tried at Wylam, it went in pieces, to the danger and alarm of the bystanders. Still desirous of succeeding, the enterprising coalowner allowed a locomotive to be made by his own engine-wright, Jonathan Foster, and such workmen as were available, under the superintendence of the viewer of the colliery, William Hedley. This engine was more successful than its predecessors, but it was too heavy for the tram-road, and it was continually getting out of repair, so that horses had usually to follow it in anticipation of its breaking down. In spite of the comparative failure of the third locomotive, Mr. Blackett was so satisfied that Hedley was on the brink of success in his experiments that he ordered a fourth engine to be constructed, and it was completed at his own engine-shop at Wylam, and placed upon the waggon-way, where it did good service for many years. After a career of usefulness, during which it was visited and scrutinised by the greatest mechanical engineers of the age, William Hedley's locomotive was sent to South Kensington, where it is exhibited in the Museum as the patriarch of the great Puffing-Billy family.

Soon after his appointment as engine-wright, the attention of George Stephenson was directed to the more efficient and economical transport of coal from the collieries to the shipping staiths. Having made various minor improvements in his machinery, he also succeeded in reducing the working expenses at one of the Killingworth pits by utilising the surplus power of an underground pumping-engine in drawing the coal to bank. But the laden waggons had thence to be hauled by horses for a distance of six or seven miles, and the price of corn being at that time exceedingly high, the expenditure entailed in the transit was proportionably burdensome. To remedy this state of matters George devoted his energies, and the first practical result of his efforts to obtain a less costly motive power on the tram-road was the application of the principle of the Inclined Plane wherever that was practicable. By thus altering certain portions of the waggon-way no doubt some saving was secured; but the slight pecuniary advantage thereby effected only served to whet his desire after a [illegible] tangible attainment in the direction of economy.

At various times George visited Wylam for the purpose of inquiring as to the locomotive experiments there carried on, and the prospects of success that might be entertained by his friend Jonathan Foster. When the engines were placed upon the tram-road, and while at work, he inspected them minutely: noting their excellencies and defects, and determining the points in their gearing that admitted of improvement. He also examined the Blenkinsop engine on the day of its being started at Coxlodge; and expressed an opinion adverse to its utility, which was shortly afterwards verified by the explosion of the boiler. The partial success attained by Hedley at Wylam, however, caused Stephenson to apply himself with ardour to the subject of locomotive improvement; and he speedily brought the matter under the notice of the Killingworth owners. From the improvements in the fixed-engines which he had already been able to effect, a favourable impression regarding their engineer's ability had been formed by Lord Ravensworth and the other lessees; and his lordship, after hearing a statement from Stephenson as to the likelihood of the proposed "travelling engine" proving a saving, authorised him to proceed with its construction without delay. Thus empowered, George commenced his task; and, working with such tools and assistance as he could command, the first Killingworth locomotive was completed in about ten months, and placed upon the waggon-way.

Stephenson's first locomotive was popularly known as "Blutcher;" and although it was exceedingly uncouth and cumbrous in appearance, as well as unsteady in its action, was a decided advancement upon anything of the kind which had previously been built. Being without springs, it jolted and jerked along in a fashion that was not calculated to give the beholder at first a favourable impression regarding its powers. The steam, after passing through the cylinders, escaped with a horrible noise, which caused the colliery-owners to be threatened with law proceedings for the terror produced by the awful machine upon cattle and horses. Besides, the economical side of the question had to be studied, and this gave only a slight advantage to the engine when compared with the horse-power which it was intended to supplant. For a time, therefore, it almost seemed as if the fate of the locomotive hung in the balance; for the interest in its success was not confined to Killingworth, and a widespread feeling of repugnance to the innovation had been generated in various parts of the country. But Stephenson was equal to the emergency. He so altered the engine that the noise created by the emission of its steam was considerably lessened; while the means which he adopted to that end increased the combustion in the furnace, and more than doubled the effective power of the machine. The adoption by the ingenious constructor of the Killingworth locomotive of what is known as the steam-blast in all probability saved his engine from condign destruction, and furthered the extension of locomotive haulage generally, not only upon private coal-carrying lines, but eventually to the public railways of the kingdom.

CHAPTER VI.

THE RIVAL SAFETY-LAMPS.

COLLIERY explosions, the harbingers of destruction to mining property, and bereavement to mining households, were familiar to the owners and workers at Killingworth. Shortly after Stephenson's removal to West Moor an explosion of fire-damp occasioned the loss of ten lives, while he was engaged in his duties as brakesman at the top of the shaft; and the calamity produced in him a painful impression, which was revived in the year 1809, when a similar catastrophe at the same pit deprived twelve persons of their existence. Other disasters of a like nature had also occurred at neighbouring collieries, and with even more fatal results; which circumstances exerted a powerful influence over the sympathetic nature of George, and impelled him to strive earnestly to discover a remedy for an evil which rendered the lives of the pitmen the price that was but too frequently paid for the acquisition of coal. It was not, however, until his advancement to the post of engine-wright that he was able to give practical effect to the generous resolution which he had formed.

Having personally to superintend the hauling of the coal over planes underground, he was often brought in contact with fire-damp. Various methods had been adopted for neutralising the foul air, caused by the frequent presence of carburetted hydrogen in certain parts of the Killingworth mine. Some of the more dangerous galleries had been built up, in order to shut off the deadly gas from the other workings. Still, the sudden escape of a "blower" might occur at any moment, even in what were considered the safest places, and the flames of the lamps or candles being then unguarded, the miners were continually exposed to the chance of mutilation or a terrible death. It was to lessen this danger to miners, and reduce the mortality among them from this cause, that Stephenson set about the invention of a lamp that would at once give sufficient light to the men while at work and prevent the inflammable gas from igniting at the flame. Mr. Samuel Smiles, in his most interesting *Lives of the Engineers*, gives an anecdote which was related to him by an old Killingworth miner, and it is characteristic of the indomitable courage of the inventor of the "Geordy Safety Lamp." Mr. Smiles says:—"One day, in 1814, a workman hurried into Stephenson's cottage with the startling information that the deepest main of the colliery was on fire! He immediately hastened to the pit-head, about a hundred yards off, whither the women and children of the colliery were running, with wildness and terror depicted in every face. In a commanding voice Stephenson ordered the engineman to lower him down the shaft in the corve. There was peril, it might be death, before him, but he would go. He was soon at the bottom, and in the midst of the men, who were paralysed by the danger which threatened the lives of all in the pit. Leaping from the corve on its touching the ground, he called out, 'Are there six men among you who have courage to follow me? If so, come, and we will put the fire out!' The Killingworth pitmen had the most perfect confidence in their engine-wright, and they readily volunteered to follow him. Silence succeeded the frantic tumult of the previous minute, and the men set to work with a will. In every mine, bricks, mortar, and tools enough are at hand, and by Stephenson's direction the materials were forthwith carried to the required spot, where, in a very short time, a wall was raised at the entrance to the main; he himself taking the most active part in the work. The atmospheric air was by this means excluded, the fire was extinguished, the people were saved from death, and the mine was preserved."

For some time previous to this incident the miners had found occasion to expostulate with Stephenson about what they considered to be his foolhardy experiments with the fire-damp. It was afterwards not an unusual thing for the men to discover him in the act of holding a light dangerously near to a fissure in the coal from which gas was escaping. As he was but a rude sort of chemist, he did not scruple to use the natural laboratory which was at hand, nor to conduct his experiments in a very unscientific fashion. His patient application to the subject, however, led him to the conclusion that by constructing a lamp with a chimney, through which a current of burnt air would be continually ascending from the flame, the inflammable gas would be prevented from igniting at the lamp, and the colliers might then pursue their labour, even in the most dangerous parts, without fear of accident. Having satisfied himself *about* the theory of his safety-lamp—although strict *his theory* was an incorrect one—he straightway so his friend, Mr. Nicholas Wood, the viewer of the colliery, to whom he gave a description of the invention, which he desired to see practically tested without delay. Mr. Wood executed drawings according to the inventor's explanation; and, armed with these, George made his way to Newcastle, where he ordered a lamp to be made by a tinsmith after the detailed design, and with a glass chimney which was to be furnished by a local glass firm. On the 21st of October 1815, the first "Geordy" lamp was tested in a part of the Killingworth mine that was highly explosive.

Accompanied by Nicholas Wood and John Moodie—the latter a man of great experience in coal-mining and knowledge of fire-damp—George descended the shaft with his lamp, and proceeded to one of the galleries into which the gas was coming with a hissing noise, from a fissure in the roof. In order to make the test as searching as possible, a deal partition was erected for the purpose of intercepting and collecting the gas, and rendering it more liable to ignition. That Wood and Moodie considered the danger was great to which George exposed himself in venturing into such a neighbourhood with an untried lamp, was shown by both retiring to a place of safety while the inventor tried the efficacy of his apparatus alone. Confident in its power to protect the property of his employers as well as the lives of his friends and himself, he boldly advanced within the dangerous precincts, and held his lighted lamp within a few inches of the fissure, and in the full current of the inflammable gas. The flame was extinguished by the deadly "blower," but without exploding the foul air which surrounded the lamp. Stephenson, although neither a philosopher nor a chemist, had gained a scientific victory. Repeated experiments by the inventor, and those who attended him, demonstrated the security of the flame under the most unfavourable circumstances; but further improvement in the construction of the lamp being desirable in order to increase its illuminating power, George, with his usual energy, applied himself to the full accomplishment of his purpose.

In all his researches into the qualities and effects of fire-damp, he had been unassisted by reference to the opinions of others on the subject. It should also be borne in mind that at the time he was prosecuting these investigations, the claims of the locomotive, then upon its trial, absorbed much of the attention and leisure at his disposal, after fulfilling his duties at the various collieries of which he was the engineer. The success of his invention so far had not been due either to the scientific accuracy of the theory upon which it had been planned, or the experiments of others in the same direction. The end in view, namely the production of a safe light-giving agent in the dark recesses of a collier's daily workshop, had been practically attained by the untiring industry and courage of the inventor. In seeking to improve his lamp, Stephenson's self-reliance had to be again called mainly into requisition, although the confidence of his friend, Nicholas Wood, had now been secured, and that gentleman rendered what assistance he could in the further experiments that were made.

When giving his evidence before the Committee on Mining Accidents, in the year 1835, George Stephenson thus referred to the tests which he instituted in order to perfect the efficiency of his lamp. He said:—"I made several experiments as to the velocity required in tubes of different diameters, to prevent explosion from fire-damp. We made the mixtures in all proportions of light carburetted hydrogen with atmospheric air in the receiver, and we found by the experiments that when a current of the most explosive mixture that we could make was forced up a tube four-tenths of an inch in diameter, the necessary current was nine inches in a second to prevent its coming down that tube. These experiments were repeated several times. We had two or three 'blows up' in making the experiments, by the flame getting down into the receiver, though we had a piece of very fine wire-gauze put at the bottom of the pipe, between the receiver and the pipe through which we were forcing the current. In one of these experiments I was watching the flame in the tube, my son was taking the vibrations of the pendulum of the clock, and Mr. Wood was attending to give me the column of water as I called for it, to keep the current up to a certain point. As I saw the flame descending in the tube I called for more water, and Wood unfortunately turned the cock the wrong way; the current ceased, the flame went down the tube, and all our implements were blown to pieces, which at the time we were not very able to replace." It was after repeated tests had been conducted, under similarly discouraging circumstances, that the "Geordy" lamp attained its great efficiency, and the high estimation in which it has been held by those for whose personal safety it was devised by the Killingworth engineer.

While the humble overseer of colliery machinery was thus endeavouring to improve his invention, one of the greatest philosophers of the age was employing his vast knowledge, experience, and resources in designing a similar contrivance. At the invitation of an influential committee which had been formed for the purpose of devising the best means of preventing fire-damp explosions, Sir Humphrey

Davy had determined to solve the problem if possible, and in connection with his investigations, the distinguished chemist visited some of the Tyneside collieries on the 24th of August 1815. Of that visit and its purport Stephenson had no information at the time; and it was not until the 9th of November following that Sir Humphrey Davy read his dissertation, "On the Fire-Damp of Coal Mines, and on Methods of Lighting the Mine so as to Prevent its Explosion," before the Royal Society of London. In the course of his experiments Davy had made the important discovery that fire-damp explosion could not pass through tubes of a certain diameter, and on that theory he invented his lamp and described it to the members of the Royal Society. But George Stephenson had proved by a practical test in one of the most dangerous galleries of the Killingworth pit, and at the hazard of his life, the same scientific fact, although in his experiment he had endeavoured to find the source of safety from explosion in another cause. Thus the two inventors were distinct but similar, and it was only when the question of priority was raised that we find any attempt to detract from the merits of either.

The first "Geordy" lamp was proved as to its safety, as we have seen, on the 21st of October 1815. Fourteen days after—on the 4th of November—Stephenson's second lamp was similarly tested, and found to be not only as safe as the first, but also of greater illuminating power. On the 9th of November the first "Davy" lamp was shown in London. In point of priority, therefore, history must give the place of honour to the "Geordy," and acquit its ingenious contriver from any charge of attempting to appropriate to himself a merit to which he was not honestly entitled; especially when we are assured that neither on the 20th of

November, when Stephenson went to order his third lamp from a Newcastle plumber, nor ten days afterwards, when the new lamp was experimented with in the pit at Killingworth, had either the inventor or his friend Mr. Wood heard of the "Davy" lamp, or of the experiments which had led to its construction.

Having been urged to bring his invention under the notice of the Literary and Philosophical Society of Newcastle-upon-Tyne, Stephenson consented, but with considerable reluctance, and upon the understanding that Nicholas Wood was to act in the capacity of spokesman upon the occasion. Upwards of eighty gentlemen having met on the 5th of December 1815, for the purpose of seeing the lamp and hearing it described, Mr. Wood proceeded to carry out the duty he had undertaken. He had not, however, a sufficient knowledge of the apparatus to enable him to answer correctly some of the questions that were put by members of the Society; and consequently George was compelled to throw off his natural diffidence and reserve, and take the matter of expounding the principle and construction of his lamp into his own hands. Having concluded his description, even to minute details, he then endeavoured to give ocular demonstration of the success which he had achieved, by various and repeated experiments with carburetted hydrogen gas, which had been collected into bladders in the mine for that purpose. His broad Northumbrian dialect, his earnestness and confidence in his invention, as well as the unassuming manner of his address, arrested attention and made a favourable impression upon his auditory; while the practical proofs which he gave of the *safety* of his invention created an animated interest among many of the gentlemen who witnessed the experiments.

The rival safety-lamps thus having been placed before the public, considerable controversy arose as to their respective merits; while some of the gentlemen who had seen the "Geordy" exhibited before the Newcastle Society declared the "Davy," upon its first appearance in the North, to be the same as Stephenson's. Sir Humphrey's friends, with more zeal than discretion, sneered at the mean condition of the Killingworth engineer, and derided his lack of chemical knowledge and consequent ability to devise such an important invention. Experience has proved, however, that much needless anger and vituperation were imported into the discussion. After the lapse of sixty-five years, we find both lamps doing effective service in lessening the danger of explosion to the miner, and both also bearing out the opinion of disinterested critics regarding them, given when the controversy was at its height.

In the year 1816 steps were taken to raise a subscription for the purpose of presenting a suitable testimonial to *the inventor of the safety-lamp.* Finding that Stephenson's claim to priority in the invention was likely to be ignored, and his labours unrewarded, his friends determined upon making an effort to secure for him that consideration to which he was entitled. It is pleasing to note that throughout the further proceedings no attempt was made, either by George or those who acted on his behalf, to detract from the honour which was due to Sir Humphrey as *an inventor of a safety-lamp.* All that was asked for, in Stephenson's interests, was the admission that his invention had been produced *before* Davy's, and that the two designs should be judged relatively by results. After a number of meetings, and much discussion on the subject, the sum of two thousand *pounds was* presented to Sir Humphrey Davy as the inventor while a purse of one hundred guineas was voted to George Stephenson for what he had accomplished towards the same end. The result was satisfactory neither to the Killingworth engineer nor to those who had espoused his cause; so, to put the matter fairly before the public, Stephenson was advised to publish a statement in which his claims should be justified by the forceful logic of facts.

With the assistance of his son, George prepared a written summary of the work in connection with his invention. That paper was the first he had been called upon to indite with a view to publication. As might be expected, the author approached his task with much misgiving; and it was only finished after several evenings had been occupied in the composition. Having been completed, corrected again and again, and copied by Robert in his best hand, the carefully-prepared document was submitted to Mr. Brandling of Gosforth, the gentleman who had suggested the desirability of its publication, who revised it so as to make it more presentable in type; and it appeared shortly afterwards in the local newspapers. The publication of Stephenson's narrative, backed as it was by the earnest advocacy of many friends who were satisfied that he had established the validity of his claim, had the effect of arousing public sympathy in his favour, and resulted in the subscription of one thousand pounds as a reward "for the valuable service he had thus rendered to mankind." At a public dinner in the Newcastle Assembly Rooms, George was presented with a silver tankard and the balance of the handsome sum which had been collected. The following inscription was engraved upon the tankard:—"This piece of plate, purchased with a part of the sum of £1000, a subscription raised for the remuneration of Mr. George Stephenson for having discovered the fact that inflamed fire-damp will not pass through tubes and apertures of small dimensions, and having been *the first* to apply that principle in the construction of a safety-lamp calculated for the preservation of human life in situations formerly of the greatest danger, was presented to him at a general meeting of the subscribers, Charles John Brandling, Esq., in the chair, 12th January 1818."

Although justice had been thus tardily extended to the Killingworth inventor, it remained for time, and the test of a prolonged use of the "Geordy" lamp, to establish for it a superiority greater by far than anything involved in the mere fact of its priority of production. Nor in expressing an opinion favourable to Stephenson rather than to Sir Humphrey, when the merits of their respective inventions come to be weighed in regard to the maximum of *safety* afforded by each, must we be considered as seeking to undervalue Davy's invention in order thereby to place an additional laurel in Stephenson's wreath. The circumstance under which the great philosopher undertook his task, and the generous relinquishment of any pecuniary interest in it after it had been completed, redound to the honour of his memory. He had been invited, by those who had great interests at stake in mining operations, to devise a safeguard against the destruction of property as well as the sacrifice of life; and his invention, as a saviour of capital, had a monetary value as well as a life-saving excellence. But, for the sake of humanity, he repudiated for himself all personal advantage, and thereby enhanced, if that were possible, the worth of his invention.

Lord Brougham, in his *Lives of Philosophers*, thus alludes to the safety-lamp which had been given to the country by his friend Sir Humphrey Davy :—

"The dreadful ravages made on human life by fire-damp explosions—that is, the burning of hydrogen gas in mines—had often attracted the notice of both the mine-owner and the philanthropist. Various inventions had been fallen upon to give light in those recesses of the earth, with so low a degree of heat as should be insufficient to explode the gas. One of them was a series of flints playing by machinery against each other, so as to give a dim light; but this had very little success: it was clumsy, and it was not effectual so as to cause its use by miners. The ventilation of the galleries by furnaces, and even by air-pumps, was chiefly relied on as a preventive; but the gas would still collect in spite of all preventives, and the destruction of a hundred or more lives was not an unusual calamity. Davy, about the year 1815, turned his attention to the subject, and after fully ascertaining that carburetted hydrogen is the cause of the fire-damp, and finding in what proportions it must be mixed with air in order to explode (between six and fourteen times its bulk), he was surprised to observe, in the course of his experiments made for the purpose of ascertaining how the inflammation takes place, that the flames will not pass through tubes of a certain length or smallness of bore. He then found that if the length be diminished, and the bore also reduced, the flames will not pass; and he further found that by multiplying the number of the tubes, their length may safely be diminished to hardly anything, provided their bore be proportionably lessened. Hence it appeared that gauze of wire, whose meshes was only one-twenty-second of an inch diameter, stopped the flame and prevented the explosion. The candle or lamp being wrapped in such gauze, and all access to the external air prevented except through the meshes, it is found that the lamp may be safely introduced into a gallery filled with fire-damp; a feeble blue flame will take place inside the gauze, but no explosion, even if the wire be heated nearly red. The theory is, but it seems very questionable, that the conducting power of the wire carrying off the heat prevents a sufficient quantity reaching the explosive compound. Subsequent inquiries seem to prove that although in a still atmosphere of explosive gas the lamp is a perfect protection, yet it does not prevent a current of gas from penetrating to the flame and exploding. It is attempted to guard against this by interposing a tin shield or screen; but a current very often in mining operations arises before any notice can be given. Had Davy's life and health been prolonged, he might have further improved his invention so as to meet this objection. He certainly never was fully convinced of its force, as I know from having discussed the subject with him; and no doubt the testimony of so great an engineer as the late Mr. Buddle, given before a Parliamentary Committee to whom the examination of this important subject was referred, deserves great attention. He positively affirmed that 'having seen 1000, and sometimes 1500 safety-lamps in daily use, and in all possible varieties of explosive mixtures, he had never known one solitary instance of an explosion.' As for the lamentable accidents which continue to happen, we can scarcely doubt that they originate in the dreadful carelessness of their own and of other men's lives, which seems to be engendered in those who are habitually exposed to great danger. That they themselves are the first to suffer for it, can only suppress the outward expression of the feelings which recklessness like this is fitted to produce."

When the heat of discussion had subsided, and experience enabled practical men to judge of the two inventions without undue prejudice in favour of either, the "Geordy" lamp was found to be superior to the "Davy," in that it afforded greater security to the miner while at work in highly explosive situations. The correctness of that opinion has frequently been demonstrated; as it is found that whenever the inflammability of the atmosphere in galleries is so intense as to render red-hot the wire-gauze of Sir Humphrey's lamp, the flame of a "Geordy" is at once extinguished—a fact that renders the latter peculiarly adapted for use in deep workings and other dangerous places.

Before leaving the subject of the rival safety-lamps, it may interest some readers to learn that now—in the year of the Stephenson centenary—the great value and excellence of the Killingworth engineer's invention have received further confirmation, after various experiments. Mr. Robert Reed, viewer of Felling Colliery, near Newcastle, and of long practical experience in mining, has for some time given his attention to the question of the safest and most efficient lighting of the mine under his charge: one, it may be noted, which upon at least two occasions prior to the invention of a safety-lamp gave deadly evidence of its explosive nature. In the year 1812, from an explosion of fire-damp in the pit, ninety men and boys were either burnt to death or suffocated; while in the following year twenty-two persons lost their lives from a similar cause. With such a record of disaster in the history of his mine, Mr. Reed was desirous of having the safest illuminating agent possible in his workings; and many investigations and various trials have proved beyond a doubt the decided superiority of George Stephenson's safety-lamp, not only over its old rival the "Davy," but also more modern contrivances. Orders have therefore been lately given for the manufacture of safety-lamps for Felling Colliery after the pattern of the "Original Geordy."

CHAPTER VII.

THE FIRST PUBLIC RAILWAY.

SATISFIED with the alterations which he had effected in his first travelling engine, but anxious to place the question of the utility of steam locomotion in the transport of coal beyond the region of uncertainty, Stephenson engaged in further inquiry and trials. To the prosecution of these he devoted much patient thought and unflagging industry. He was sensibly aware that his engine had many defects; not the least of which was its somewhat complicated gearing. Directing his efforts, therefore, to the construction of another which should possess in its form and cost the desiderata of simplicity and economy, he so far succeeded as to be able to take out a patent, in February 1815, for a locomotive which combined these essentials in a remarkable degree. It contained the germ of much that has been subsequently produced, and may be considered as the type upon which has been moulded the locomotive of the present day.

But Stephenson did not confine his attention alone to the

improvement of the motive-power which he desired to see adopted. In his experiments he had arrived at the settled conviction that a superior form of travelling engine demanded a better road upon which to run than any that had been hitherto laid down. In the laying of rails little care had been taken to remove inequalities in the roadway. The levelling had been very imperfect at the outset, and the lack of systematic repairs allowed the rails to become so depressed at intervals as to cause not only great loss of power in the engine, but considerable detriment also to the machinery, through the continued jolting which it sustained in the course of the journeys. To alter this state of matters he applied himself with energy, and the result of his thought and labour appeared in an improved rail and chair for the road, and in malleable iron wheels and a steam substitute for springs in the locomotive. These various improvements were specified in a patent granted in September 1816, jointly to William Losh, ironfounder, of the Walker Iron Works, Newcastle, and George Stephenson—Mr. Losh furnishing the capital.

With an improved locomotive and railway, Stephenson was now enabled to conduct the coal traffic from his colliery with increased regularity, while the advantage of greater economy in working was found to be on the side of the engine when compared with the horse-haulage. He had therefore arrived at a point in his labours from which he might prosecute further researches into the region of possibility without being harassed as formerly with the thought that his efforts were unproductive of practical benefit to his employers. In October 1818, with Mr. Nicholas Wood, he instituted experiments with a dynamometer, constructed by himself, for the purpose of determining with accuracy the resistance offered to carriages while travelling on the railway. These observations were the first of their kind, and confirmed the deductions of Vince and others as to the theory that friction is a constant quantity at all velocities—a theory that was then scouted by many engineers. Thus was he being gradually trained and fitted for the great work that lay before him as the pioneer of the railway system.

Five years had elapsed from the completion of the first Killingworth locomotive before any attempt was made to extend the principle to the working of the coal traffic in other places. Stephenson himself had great confidence in his hypothesis that the union of the travelling engine and the railway was destined to effect stupendous results; but his locomotives continued to do their accustomed work without exciting more than a passing interest in casual observers of their immense power and ungainly appearance. The general public took no interest in an uninviting subject such as the conveyance of coal, and if other coalowners did really from time to time scrutinize the improved motive-power on the Killingworth Railway, they took care to keep the result of their inspection to themselves. So disheartened did George become at length with the apathy of his countrymen regarding the subject which mainly engrossed his thoughts, that he once more reverted to his old idea of crossing the Atlantic and pushing his fortune under less depressing and discouraging conditions.

In the year 1819, Stephenson began to reap the first-fruits of his patience, energy, and toil. His locomotives having effected a decrease in the cost as well as an increase in the speed of the traffic over the Killingworth line, an important Durham Company decided upon altering their waggon-way so as to allow the adoption of locomotive p in the transit of coal from Hetton Pit to the River Wear at Sunderland. Stephenson was asked to devise and superintend the alterations, and the manner in which he brought these to a successful termination, as well as the careful economy which guided him in his work, showed that while he was competent to deal with engineering questions in a firm and accurate manner, he was too wise to risk the prosperity of the undertaking in order that he might achieve a showy but questionable result. The length of the Hetton line was about eight miles, crossing in its course the hill called Wardon Law. In forming it, Stephenson endeavoured as much as possible to adapt the working of the railway to the natural condition of the ground over which it passed. Consequently, he utilised five of the heavy gradients as self-acting inclines, and the remaining two were worked by fixed engines. Other portions of the line being suitable for locomotive power, travelling engines conveyed the waggons to and from each incline. On the 18th November 1822 the new Hetton Railway was opened for traffic; five locomotives constructed under Stephenson's direction being upon the line under the charge of his brother Robert. With a train of seventeen waggons, weighing about sixty-four tons, each engine travelled at a speed of about four miles an hour. A large number of spectators assembled to witness the inauguration of George's first important work in railway engineering, the complete success of which gratified the proprietors of the line, and gave increased confidence to the engineer to engage in a larger and more important undertaking.

Edward Pease, the projector of the first public railway, had many difficulties to surmount ere the formation of the Stockton and Darlington Railway Company became an accomplished fact. Although the scheme was one of public utility rather than of private advantage, those who were most likely to be benefited by its completion either withheld support or offered opposition. Among the inhabitants of South Durham generally the proposed line was declared to be a preposterous idea, which, if carried out, could only end in disaster to its unfortunate shareholders. But Mr. Pease held firmly to his purpose, and by the aid of relatives and personal friends the required capital was subscribed. It is said that the shares allotted in the town of Stockton did not amount to twenty.

In the year 1818, the first application was made to Parliament for an Act authorising the construction of the contemplated line, but the Bill was successfully opposed by the Duke of Cleveland. Another survey having been made by which the fox-covers of his Grace were not invaded, the promoters were more fortunate in their second application for Parliamentary powers, and the first Stockton and Darlington Railway Act received the Royal Assent on the 19th April 1821. By the Act, authority was given "for making and maintaining a railway or tramroad from the river Tees, at Stockton, to Witton Park Colliery, with several branches therefrom, all in the county of Durham;" but no power had been asked for working the proposed railway by locomotives, and it was not until the year 1823 that the omission was rectified by a clause in a supplementary Act passed on the 23rd of May. In that second statute, the company was empowered to "make, erect, and set-up one permanent or fixed steam-engine, or other proper machine, in such convenient situation as might be selected; also to use locomotives or movable engines for the purpose of facilitating the transport, conveyance, and carriage of goods, merchandise, and other articles and things, upon and along the same roads, and for the conveyance of passengers upon and along the same roads." The use of locomotives on the first public railway was urgently recommended by the engineer, George Stephenson, and the necessary power to employ such haulage was obtained on his advice.

The construction of the Hetton Railway, there can be little doubt, was the means of bringing the capabilities of the engineer before the Darlington projector in a prominent manner. The first interview between Edward Pease and George Stephenson took place ten days after the passing of the Stockton and Darlington Act of 1821, and was referred to by Mr. Nicholas Wood, who was also present upon the occasion, in an address given by him many years after. Mr. Wood said:—"The fact is, we rode on horseback from Killingworth to Newcastle, a distance of five miles, travelled from thence by coach, thirty-two miles, to Stockton, then walked along the proposed line of railway, twelve miles, from Stockton to Darlington. We had then the interview with Mr. Pease, by appointment, and afterwards walked eighteen long miles to Durham, within three miles of which I broke down . . . . but was obliged to proceed, the beds being all engaged at the 'Travellers' Rest.' This interview with Mr. Pease, which was on the 19th of April 1821, had the effect of Stephenson being ultimately appointed engineer to the Stockton and Darlington Railway."

At various times prior to 1818, ineffectual attempts had been made to develop the mineral resources of South Durham by means of a canal; and in that year the scheme was revived, in opposition to the railway project of Edward Pease. On economical grounds the railway was found to offer greater advantages, and that gentleman therefore gave all the weight of his great influence and advocacy in favour of the iron and against the water highway. The latter proposal being abandoned by its supporters, Mr. Pease and his friends were left to carry out the construction of the railway, while their opponents looked on their efforts with chilling apathy or heated disapproval.

The first survey of the projected line was conducted by Mr. George Overton, an engineer and contractor of South Wales, who felt considerably irritated at the rejection of the Bill of 1818, and at the fact of his recommendations having been set aside by Mr. Pease and the committee appointed to carry the scheme to completion. But the projectors foresaw the difficulties and expense that would be encountered if they persisted in prosecuting a route which brought them into antagonism with such a powerful landowner as the Duke of Cleveland, and they wisely determined to have a second survey taken with a deviation from the line of the first, which they considered was calculated to allay the irascibility of the noble fox-preserver. Mr. Overton accordingly made another survey, and by dint of most strenuous exertions, for the purpose of enlisting the support of members of Parliament and removing the objections of peers of the realm in regard to the project, the second Bill was carried, as we have stated, and the committee proceeded to the practical portion of their labours.

Immediately after the visit to Mr. Pease of Stephenson and his friend Nicholas Wood, the directors of the railway took into consideration the appointment of an engineer for the line. Satisfied of the sterling character of the north-country candidate for the appointment, the directors empowered Edward Pease to write to George asking him for information as to his terms for re-surveying the route as laid out by George Overton, with a view to ascertaining whether the construction of such a line would be practicable; and, generally, to determine the possibility of effecting greater economy and utility in the construction and working of the railway than would be attained by a strict adherence to Overton's survey. The letter written to Stephenson by Mr. Pease was a model communication, exhibiting careful forethought and business capacity in the writer, and a true estimate of the capability and trustiness of him to whom it was addressed. The reply of the engineer was as follows:—

"EDWARD PEASE, ESQ.

"SIR,—After carefully examining your favour, I find it impossible to form an accurate idea of what such a survey would cost, as not only the old line must be gone over, but all the other deviating parts, which will be equal to a double survey, and, indeed, it must be done in a very different manner from your former one, so as to enable me to make a correct measurement of all the *cuts and batteries* on the whole line. It would, I think, occupy me at least five weeks. My charge shall include all necessary assistance for the accomplishment of the survey, estimates of the expense of cuts and batteries on the different projected lines, together with all remarks, reports, &c., of the same. Also the comparative cost of malleable iron and cast iron rails, winning and preparing the blocks of stone, and all materials wanted to complete the line. I could not do this for less than £140, allowing me to be moderately paid. I assure you, in completing the undertaking, I will act with that economy which would influence me if the whole of this work was my own.

"GEORGE STEPHENSON.

"KILLINGWORTH COLLIERY, *August 2nd*, 1821."

Arrangements having been concluded, Stephenson began his survey in the autumn of 1821, and the work was completed with so much satisfaction and credit to the directors and himself, that he was appointed engineer to the company, at a salary of £660 per annum, inclusive of the cost of assistance and personal expenses. In the labour of laying out the railway he was aided by his son Robert, and a young man named John Dixon who eventually attained the position of consulting engineer to the directors. As chairman of the company, Mr. Thomas Meynell of Yarm laid the first rail, and the ceremony was made the occasion of public rejoicing. This preliminary observance took place on the 23rd May 1822, in the vicinity of St. John's Well, Stockton.

The formation of the line having been fairly started, Stephenson next proceeded, but with cautious circumspection, to advocate the use of locomotives where practicable; one portion of the railway going over a hilly tract of country which involved the construction of a heavy incline, for the working of which travelling engines would have been unsuitable. In making his first estimate he refrained from setting down any sum for locomotive charges, seeing that up to that time the directors had contemplated the employment of horse-power only for the more even parts of the road. He now began to see, however, that the adoption of locomotive haulage depended upon his being able to satisfy the directors that such a motive power was not only more economical, but at the same time that it would conduce to greater regularity in the working of the line. For the purpose of having the question decided, Stephenson waited upon Mr. Edward Pease, whom he invited to Killingworth, in order that the locomotives might be there seen at work, and that Mr. Pease might inform himself as to the cost of such haulage from the colliery accounts. Benjamin Thompson, of Eighton Banks, having procured a patent for a fixed engine, and urged its employment on the *Stockton* and Darlington line, Mr. Pease was natura uncertain as to which description of engine should be secured. However, having great confidence in the value of his own engineer's opinion, and for the purpose of testing the powers of the Killingworth engines, he agreed to visit the Northumberland Colliery before definitely advising his brother directors in the matter.

On the day of his arrival at Killingworth, Mr. Pease found that George was engaged in one of his acts of duty down the pit, but being called therefrom, the engineer was quickly employed in giving his visitor a practical illustration of the wonderful capabilities of the locomotive. Thenceforward the railway projector associated the question of the perfect success of his scheme with the union of the travelling engine and the rail. The result of the visit to Killingworth led the directors to decide that locomotives should be employed on the Stockton and Darlington Railway.

The upright character of George Stephenson was exhibited when the question arose as to the kind of rails which should be laid upon the road. As already stated, he was interested jointly with Mr. Losh, of Walker, in a rail which had been patented; and its adoption would have conferred upon himself considerable pecuniary benefit. He preferred, however, to act conscientiously in the matter, and when asked for his opinion, strongly advised the use of malleable rails in preference to cast-iron ones, in the manufacture and sale of which he had a personal interest. The directors gave orders for the laying of the former, as an experiment, only on a portion of the line.

Before the projection of the Stockton and Darlington Railway, considerable difficulty had been experienced in the building of locomotives, on account of the want of skilled mechanics for the work. The whole of Stephenson's engines for the Killingworth and Hetton lines had been made by colliery operatives, and it was felt that before much improvement could be effected in the form or action of the locomotive, workmen would be required who were specially qualified for such labour by having undergone a training in engine-fitting. Throughout his experiments in connection with the construction and alteration of his engines, George had been harassed, and his efforts had been retarded, by reason of the men who were employed to assist him being frequently unable to carry out his instructions; and for the purpose of remedying this hindrance, as well as engaging in a profitable speculation, he determined to commence business as a manufacturer of locomotives. He had intact the amount of his testimonial for the invention of a safety-lamp; and with that sum, and £1000 put into the concern by Edward Pease and his cousin Thomas Richardson—the latter a gentleman whose influence had been of great value in projecting the Stockton and Darlington Railway—the well-known engine works in Forth Street, Newcastle, were founded, under the firm of Robert Stephenson & Company.

On the 16th September 1824, the first order for locomotives was given to the new firm by the directors of the line; and in obedience to that order, two engines were built at a charge of £500 each. The first to be employed on a public railway was named "Locomotion," and it headed the inaugural procession on the opening day. After serving well its generation, it now stands upon a pedestal at the Darlington Station, in North Road, where it was placed in June 1857. Besides furnishing locomotives for the line, R. Stephenson & Co. erected at the Brusselton hill-top two thirty horse-power stationary engines, with combined axle, for drawing trains up the incline. Two engines of similar construction, but of half that nominal power, were erected at Etherley hill-top also by the same firm. The cost of these four engines was £5465, 10s.; the size of the working cylinders being thirty inches for the two at Brusselton, and those at Etherley having cylinders of twenty-two inches.

"The conveyance of passengers," Mr. Wood remarks, "did not form a part of the original intentions of the promoters. The conveyance of coals at the cheapest possible rate was the desideratum, and the principle which Stephenson was instructed to proceed upon. High rate of speed was no element for the consideration of either directors or engineers. Heavy loads, conveyed at moderate rates of speed, were alone considered. Hence the locomotive engines to be used on the Stockton and Darlington Railway were constructed to travel from four to six miles an hour, with the heaviest load which the power of the boiler in raising steam enabled them to accomplish; and hence also we find, on Messrs. Walker and Rastrick's visit in 1829, ... they place the performance of the engines at 47¾ tons of goods, 23⅓ tons weight of carriages, the engine and the tender weighing 15 tons—making altogether a gross weight of 86½ tons, moved at five miles an hour." The purpose of the promoters was to encourage the use of the railroad, on certain conditions, by the owners of waggons and coaches plying between the terminal towns. In 1833 the company became the sole carriers of goods and passengers.

The main line being completed, the directors made arrangements for the opening thereof for traffic. In announcing the preliminary ceremony, the following handbill was issued:—

THE

STOCKTON & DARLINGTON

RAILWAY COMPANY

Hereby Give Notice,

THAT the FORMAL OPENING of their RAILWAY will take place on the 27th instant, as announced in the public Papers.—The Proprietors will assemble at the Permanent Steam-Engine, situated below BRUSSELTON TOWER,* *about nine miles west of* DARLINGTON, *at eight o'clock, and, after examining their extensive inclined Planes there, will start from the Foot of the* BRUSSELTON *descending Plane, at nine o'clock, in the following Order:—*

1. THE COMPANY'S LOCOMOTIVE ENGINE.
2. The ENGINE'S TENDER, with Water and Coals.
3. SIX WAGGONS, laden with Coals, Merchandise, &c.
4. The COMMITTEE, and other PROPRIETORS, in the COACH belonging to the COMPANY.
5. SIX WAGGONS, with Seats reserved for STRANGERS.
6. FOURTEEN WAGGONS, for the Conveyance of Workmen and others.

☞ *The WHOLE of the above to proceed to STOCKTON.*

7. SIX WAGGONS, laden with Coals, to leave the Procession at the DARLINGTON BRANCH.
8. SIX WAGGONS, drawn by Horses, for Workmen and others.
9. Ditto. Ditto.
10. Ditto. Ditto.
11. Ditto. Ditto.

The COMPANY'S WORKMEN to leave the Procession at DARLINGTON, and DINE at that Place at ONE o'clock; excepting those to whom Tickets are specially given for YARM, and for whom Conveyances will be provided, on their arrival at STOCKTON.

TICKETS will be given to the Workmen who are to dine at DARLINGTON, specifying the Houses of Entertainment.

The PROPRIETORS, and such of the NOBILITY and GENTRY as may honour them with their company, will DINE precisely at THREE o'clock, at the TOWN HALL, STOCKTON.—Such of the party as may incline to return to DARLINGTON that Evening, will find Conveyances in waiting for their Accommodation, to start from the COMPANY'S WHARF there precisely at SEVEN o'clock.

The COMPANY take this Opportunity of enjoining on all their WORKPEOPLE that attention to *Sobriety* and *Decorum*, which they have hitherto had the Pleasure of observing.

The COMMITTEE give this PUBLIC NOTICE, that all Persons who shall ride upon, or by the sides of, the RAILWAY, on Horseback, will incur the Penalties imposed by the Acts of Parliament passed relative to this RAILWAY.

RAILWAY OFFICE, *Sept. 19th, 1825.*

---

* Any Individuals desirous of seeing the Train of Waggons descending the inclined Plane from ETHERLEY, and in Progress to BRUSSELTON, may have an Opportunity of so doing, by being on the RAILWAY at ST HELEN'S AUCKLAND not later than Half-past Seven o'clock.

The opening of the first public railway was effected with every expression of satisfaction and rejoicing on the part of the promoters and the public in general. The prophets of evil regarding it were obliged to conceal disappointment at the success of the scheme, and wait for other opportunities of exhibiting their rancorous feelings. On the 27th of September, 1825, the committee met, in terms of the programme, at the bottom of Brusselton incline, after inspecting the Etherley fixed engine. The train, loaded with coals and goods, was then drawn up the eastern ridge, a distance of one thousand nine hundred and sixty yards, by the Brusselton engine, in seven and a half minutes; when it was lowered on the incline at the east side of the hill, a further distance of eight hundred and eighty yards, in five minutes. At the foot of the plane, "Locomotion" stood ready to be attached to the carriages, amid the wonderment, fear, and admiration of the assembled spectators. As well to give effect to the procession as for the purpose of guarding against accident, men were employed to ride in front of the engine and herald its approach, by the exercise of their voices and the waving of flags.

The first public railway train consisted of thirty-eight carriages, and its engine was driven by George Stephenson, who was both the designer and builder. It had been intended to limit the number of passengers to three hundred, but the great pressure of the crowd upset the arrangements in that particular, and by the time the train reached Stockton, not less than six hundred persons occupied seats or hung on the vehicles in one position or another. The distance of twelve miles, with the various stoppages upon the road, was accomplished in three hours and seven minutes. All along the route sightseers were crowded on every bit of vantage ground, and when the procession neared Stockton it was followed by a large number of persons on foot, in vehicles, and on horseback. It has been said that "the passengers by the engine had the pleasure of observing the difference between the engine, with her six hundred passengers and load (of eighty tons), and the coach, with four horses and only sixteen passengers." At the close of the day the good performance of "Locomotion" formed the subject of general conversation as well as after-dinner speeches.

The Stockton and Darlington Railway having been constructed for the purpose of developing the coal trade of South Durham, and bringing the supply from the pits of the district into the export market, it was satisfactory to the directors that one of the first contracts made by them for the transport of coal along the line was that for the carriage of one hundred thousand tons annually for five years by one London firm alone. That traffic of itself insured to the shareholders a dividend of four per cent. on the cost of construction, while another immediate effect of the opening of the railway was the reduction of the former price of coal to the extent of one-third in favour of the general public of the neighbourhood.

In the month of October, 1825, the first railway passenger coach commenced running; horse-power being employed, and the directors advancing the sum of £25 to one Thomas Close, upon his giving security for the purchase of a horse and harness for the purpose. The directors do not appear to have attached much importance at first to the passenger traffic as a source of revenue; and this branch of conveyance drifted into the hands of private proprietors, who worked it on terms of lease. The first coach was built at *Newcastle, by Stephenson*, after his own design, and at the cost of the company. It resembled a showman's caravan, and it formed part of the procession at the opening of the line. The seats were placed along the sides of the interior, and a deal table stood in the centre of the vehicle, which had been named "Experiment" by its designer. It was the only passenger carriage of the company in the year 1825. In Longstaff's *History of Darlington*, the sensations attendant upon an early railway journey between Stockton and Darlington are thus described:—"The coach had no springs of any kind, and yet the motion was fully as easy as in any coach on the road. A very slight jolt is felt, accompanied with a click or rattle, every time the wheels pass over the joints of the several rails, and also at the breaks which occur at the different passing places, and then, if anything, feels harsher than in a coach. At any bends of the road, or other places where the view is obstructed, the coachman blows a horn to give warning of his approach to any waggons or vehicles that may be coming or going on the way. Some parts of the way were laid with rails of cast-iron, joined at every four feet, and in coming upon these the jerks and jolts were more frequent, more audible, and more sensible, resembling exactly . . . the clinking of a mill hopper."

The first lessee for the conveyance of passengers was a Darlington contractor, named Pickersgill, and the success of his venture soon led others to apply to the directors for "running powers," until, in the years 1831–32, we find seven different coaches, belonging to various proprietors, performing journeys at stated times, regulated by special orders from the directors. The subjoined handbill, one of the earliest relating to railway passenger traffic, is now of historical interest:—

STOCKTON & DARLINGTON RAILWAY.

THE COMPANY'S COACH

CALLED THE

EXPERIMENT,

Which commenced Travelling on MONDAY, the 10th of OCTOBER 1825, will continue to run from *Darlington* to *Stockton*, and from *Stockton* to *Darlington*, every Day [Sundays excepted], setting off from the DEPOT at each place at the times specified as under, (*viz.*) :—

ON MONDAY,

From Stockton at half-past 7 in the Morning, and will reach Darlington about half-past 9 ; the Coach will set off from the latter place on its return at 3 in the Afternoon, and reach Stockton about 5.

TUESDAY,

From Stockton at 3 in the Afternoon, and will reach Darlington about 5.

On the following Days, viz. :—

WEDNESDAY, THURSDAY & FRIDAY,

From Darlington at half-past 7 in the Morning, and will reach Stockton about half-past 9 ; the Coach will set off from the latter place on its return at 3 in the Afternoon, and reach Darlington about 5.

SATURDAY,

From Darlington at 1 in the Afternoon, and will reach Stockton about 3.

Passengers to pay 1s. each, and will be allowed a Package of not exceeding 14 lb., all above that weight to pay at the rate of 2d. per Stone extra. Carriage of small Parcels 3d. each. The Company will not be accountable for Parcels of above £5 Value, unless paid for as such.

Mr. RICHARD PICKERSGILL at his Office in Commercial Street, Darlington ; and Mr. TULLY at Stockton, will for the present receive any Parcels and Book Passengers.

The slow rate of speed of the early railway trains enabled persons whose greed was stronger than their morality sometimes to indulge in cheap if not costless rides. This was accomplished by jumping on the waggons when in motion, and leaping off the same before one or another of the stations had been reached. So frequent did these fraudulent acts become that notice was given, by placards posted along the line, that lawful penalties would be stringently enforced against offenders. It was surmised, possibly without any substantial foundation for the thought, that the engine-drivers of the coal and goods trains profited by a practice that was prejudical to the interests of the various coach proprietors, whose number had increased to seven in the year 1832. Each proprietor paid the duty then demanded by the State for the carriage of passengers, besides being required to have the ordinary license as a hackney coachman. A single horse was attached to each coach; the fares being respectively one penny for an outside passenger, and threehalfpence for an inside, per mile. These private owners conducted the passenger traffic until the year 1833, when the company bought up the rolling stock from the various parties, and thenceforward had the conveyance of travellers in its own hands. It is a fact worthy of notice that passengers were regularly carried by these coaches over a total length of 300,000 miles, and without injury to life or limb, during the seven years immediately following the opening of the first public railway.

CHAPTER VIII.

THE RAILWAY ENGINEER.

BEFORE his appointment as engineer to the Stockton and Darlington Railway, and while so engaged at the Killingworth and other collieries, Stephenson watched over the education of his son with the most anxious care. After attending Mr. Bruce's school in Newcastle for about four years, during which period he gave promise of his future distinction in the larger arena of life, Robert left his schoolmaster in order to be apprenticed to Mr. Wood, and thus learn the duties of a viewer. He was so employed for nearly three years, and became acquainted with the various departments of colliery work. The knowledge gained in this way was afterwards of great service to his father, as well as of advantage to himself; and during this time the daily intercourse between father and son, while it served to quicken the perceptive powers of the young man, also enabled the parent to obtain information on many points connected with the physical and other sciences of which he had previously known but little. The subject of the locomotive and its possible future engrossed, in their usual evening discussions, much of their attention and study. It is not surprising, therefore, that Robert Stephenson became, if possible, even more enthusiastic than his father regarding the prospects of the locomotive and the rail.

Believing that a thorough training in technical science was necessary to an engineer's success in the higher branches of his profession, Stephenson determined to take his son from the duties of an under-viewer, which he then fulfilled, and send him to the University of Edinburgh. This was done in October 1822, while the Stockton and Darlington line was in course of progress. After a stay of six months in the Scottish metropolis, Robert returned to Killingworth with a prize for mathematics, which he had gained by his industry and ability. While serving his apprenticeship, he had various opportunities of witnessing improvements in colliery waggon-ways, as these were being carried out under the direction of his father upon one or another of the coal-lines of Northumberland or Durham, as well as during the reconstruction of the Hetton Railway, and the making of others. It was not, however, until the surveying of the Stockton and Darlington line that the "slight, spare, bronzed boy" was called upon to take any active part in the formation of a railway.

About the year 1821, the rapid expansion of the trade of South Lancashire caused the creation of further facilities in the transit of merchandise to be seriously entertained by commercial men. The cotton mills and their produce had of late increased so much that the ordinary channels of conveyance, the waggon and the canal, were found to be inadequate for meeting the demands made upon them. Greater speed in the delivery of goods was desired, and the passing of the Stockton and Darlington Railway Act, on the 19th of April in that year, concentrated public attention in these two great centres of industry to the project of a tramroad or railway between the important shipping mart of Liverpool and the "cotton metropolis." A Liverpool merchant, Mr. Sandars, took a leading part in bringing the scheme to maturity. He instituted inquiry into the results of locomotive power, as employed on the private lines of the northern coal-field, and being satisfied therewith, he came to the conclusion that a public railway was desirable in the west, alike on utilitarian and economical grounds. His opinion was fully borne out by a treatise on the subject of a "General Iron Railway," which had been published by Thomas Gray of Nottingham about that time; the "Observations" in question having met with a large circulation, and aroused the thoughtful consideration of many people to the subject.

Having become prominently indentified with the Liverpool and Manchester Railway scheme, Mr. Sandars was associated in the work of projecting the line by Mr. William James, of West Bromwich, who had for some time taken an active interest in the laying down of tramways. At one time an iron manufacturer, engaged in a large business, James possessed considerable influence among gentlemen in the neighbourhood of Liverpool; although from having been unsuccessful in commerce, he now followed the profession of a land-agent and surveyor. The first survey of the proposed line was entrusted to him, and it was carried out with considerable difficulty, on account of adverse popular prejudices. While thus employed, it was not an unusual thing for Mr. James and his men to be met by parties of farm labourers armed with sticks, pitchforks, and even guns; prepared to dispute the passage of the surveyor and his assistants by a recourse, if necessary, to extreme violence. So fierce grew the opposition to the peaceful invaders of private grounds, that at length it was found necessary to call in the aid of a noted pugilist, whose special province it was to guard the theodolite used in the survey from destruction at the hands of an infuriated mob.

Upon two occasions Mr. James went to Killingworth, where he saw one of Stephenson's locomotives at work, and was so pleased with his inspection both of the engine and the railway, that he promised to advise the adoption of a similar road between Liverpool and Manchester, and the employment of steam locomotion in the working of the line. Satisfied with his interview with Stephenson, James returned to Liverpool, but considerable delay occurred after his first survey had been completed, and it was found that another was required to rectify imperfections in the work. In this emergency, and in order to complete the plans and estimates for embodiment in the application to Parliament during the ensuing Session, Stephenson was asked to render the needed assistance, and Robert was despatched by his father to Mr. James for that purpose. In spite of the aid thus given, the surveyor was unable to fulfil his engagement to the committee, and that body was in consequence compelled to call in the services of an engineer to finish the plans and estimates for the line. The energy which had been displayed by Stephenson in the fulfilment of his duties to the Stockton and Darlington Company having attracted the attention of Mr. Sandars, and led him also to visit Killingworth, a conversation with the Northumberland engineer confirmed the opinion which the Liverpool projector had formed regarding Stephenson's ability, both from public rumour and private information.

On his return from Killingworth, Mr. Sandars reported to the committee the result of his journey, and strongly urged that Stephenson should be appointed their engineer. Effect being given to that recommendation, the committee proceeded with their new adviser to inspect the progressional works of the Stockton and Darlington line, after which the party journeyed to Killingworth, where the locomotives were tried under various conditions; all of which tests satisfied the gentlemen present, and they determined to recommend that a company should be at once formed for carrying the scheme for a railroad between Liverpool and Manchester to completion. The first prospectus of the company was dated the 29th October, 1824, and advocated the construction of the line "as a cheap and expeditious means of conveyance for travellers;" holding out the fair prospect of a public accommodation, the magnitude and importance of which could not be immediately ascertained.

The shares in the new company having been speedily taken up, the directors and their engineer made the necessary arrangements for presenting their Bill to Parliament in the Session of 1825. Before drafting the proposed measure, however, further steps were taken to remove doubts which still existed in the minds of some of the projectors as to the practicability of using locomotives on the railway when completed. For that purpose a second visit to Killingworth was undertaken by the committee, which was soon after followed by a third. The latter journey was made in January, 1825, and the gentlemen connected with the company who formed that third deputation to the scene of George Stephenson's early mechanical successes were accompanied by consulting engineers, in order that the committee might receive advice from men of practical experience, in any future deliberations upon the question of adopting the still doubtful travelling engine as a motive power.

One of the first effects of the projection of the Liverpool and Manchester Railway was an offer of conciliatory proposals by the canal companies, who had hitherto had the monopoly of conveyance between the two towns, and who had frequently turned their exclusive position to account by charging exorbitant freights for the services which they rendered to trade and commerce. Hitherto these monopolists had ridiculed the idea of railways ever becoming serious opponents to their extortionate exactions. Now the case was altered. The Stockton and Darlington line would speedily be opened for traffic, and the utility of public railways might soon be no longer a matter of conjecture but of established fact. Under the circumstances, therefore, it was policy to endeavour to conciliate traders by lowering the canal rates, and by promising to introduce steam power in the water-carriage of merchandise. But the die had been cast; offers of conciliation came too late; and the promoters of the second public railway resolved to prosecute the scheme, until success crowned their efforts.

Finding the advocates for railway accommodation thus determined, the canal companies circulated pathetic appeals for the purpose of inciting public opposition to the proposed line. The newspaper press, too, was employed to write it down. The occupiers of houses along the intended route were assured that their property would inevitably be consumed, by being set on fire through the plentiful distribution of red hot cinders from the hateful locomotives.

Horses would become extinct were railways to extend throughout the country. The preservation of game would be impossible. Hens would refuse to lay; cows would be unable to yield their milk for the use of man; fox-hunting would be a thing of the past. There was but one element out of which the prophets of evil formed matter for congratulation to themselves: those who had the temerity to travel by rail would be blown in pieces through the bursting of boilers—always provided that the weight of the engine permitted people to travel at all!

In the face of opposition, ridicule, and abuse, George Stephenson carried out his survey. Now, he was being threatened with immersion in a duck-pond; then, his levelling operations were stopped by a number of armed gamekeepers. The agricultural classes looked upon the north-country engineer as a fiend incarnate, who had entered upon a mission of extermination for their especial harm. Some landed proprietors regarded him with feelings of unmitigated hatred. Others looked upon him as a maniac, whose condign incarceration in a lunatic asylum was alike demanded in the interests of society and the owners of estates. In general estimation he was a social Ishmael, whose hand was against every man's, and whom it was the duty of every man to oppose by means fair or foul. Writing to Mr. Joseph Pease, on October 19th, 1824, Stephenson said:—"We have sad work with Lord Derby, Lord Sefton, and Bradshaw, the great canal proprietor, whose grounds we go through with the projected railway. Their ground is blockaded on every side to prevent us getting on with the survey. Bradshaw fires guns through his grounds in the course of the night, to prevent the surveyor coming on in the dark. We are to have a grand field-day next week. The Liverpool Railway people are determined to force a survey through, if possible. Lord Sefton says he will have a hundred men against us. The Company think those great men have no right to stop a survey; it is the farmers only who have a right to complain, and by charging damages for trespass, it is all they can do." Depressed, disheartened sometimes, but with unfailing faith in the ultimate triumph of his foster-child, the locomotive, George held courageously on his rough and difficult path, till at length his survey was completed, and the Liverpool and Manchester Railway Bill was set down for argument before a committee of the House of Commons.

The counsel for the promoters of the line were Mr. Adam, Mr. Serjeant Spankie, Mr. William Brougham, and Mr. Joy. The opponents of the Bill were represented by Mr. Alderson and Mr. Parke, both of whom became Barons of the Exchequer, and by Messrs. Harrison, Erle, and other counsel of note. After a month had been occupied in taking evidence as to the difficulties experienced by traders in the conveyance of their goods to and from the towns which were to form the termini of the proposed railway, Stephenson was called as a witness for the engineering phase of the question before the Parliamentary Committee. If his statements could be shown to be devoid of probability, or were otherwise unworthy of credence, then the Bill was doomed. The odds were fearfully against the self-tutored enginewright from Northumberland as he stood within the witness-box at Westminster; for how could he hope, single-handed, and in the broad dialect of a northern village, to contend against the forensic acumen and eloquence of such distinguished London barristers as were arrayed against him? Besides, he had gone to St. Stephen's for the purpose of turning society upside down, and society had a right to object to the action—especially when the performer was a nobody who hailed from the dusky neighbourhood of a coal-pit, and who spoke an outlandish gibberish that genteel people cared not to understand!

Confronted by the sneers and ridicule of the opposing advocates, by whom he was subjected to repeated interruptions, George told the artless tale of his experience as a colliery brakesman, from the year 1803 when he removed to Killingworth. Step by step he traced his gradual rise from the position of a manual workman to that of a railway engineer. Sometimes with a struggle for utterance which gave his enemies a momentary gleam of hope that he had broken down, he recapitulated the various improvements which he had carried out upon the lines immediately under his charge; the railways which had been constructed or were in progress under his direction at Burradon, Mount Moor, Springwell, Bedlington, Hetton, and Stockton; with the engines which he had designed and built, thirty-nine of which were stationary, and sixteen locomotive. With a perceptible feeling of pride he spoke of the Killingworth line and its travelling engines, which had worked thereon with great efficiency for eleven years.

His plans for the line formed subject-matter for a cross-examination of the most searching character, for throughout his experience he had never been called upon to deal with engineering difficulties of such formidable proportions. The details of bridges, crossings, gradients, and tunnels were all used with bewildering iteration to confuse the witness and shake his credibility, by extracting from him answers to questions of which he had not caught the purport. Upon the point of speed, some members of the committee entertained views of a very cautious nature, which the counsel for the opposition were not slow to take advantage of in the interests of their clients. When years had elapsed, and railways were regarded as public necessities, Stephenson referred to his appearance at this time as a Parliamentary witness in the following terms:—"I pledged myself to the directors to attain a speed of ten miles an hour. I said I had no doubt the locomotive might be made to go much faster, but that we had better be moderate at the beginning. The directors said I was quite right; for that if, when they went to Parliament, I talked of going at a greater rate than ten miles an hour, I should *put a cross upon the concern.* It was not an easy task for me to keep the engine down to ten miles an hour, but it must be done, and I did my best. I had to place myself in that most unpleasant of all positions—the witness-box of a Parliamentary Committee. I was not long in it before I began to wish for a hole to creep out at! I could not find words to satisfy either the committee or myself. I was subjected to the cross-examination of eight or ten barristers, purposely as far as possible to bewilder me. Some member of the committee asked if I was a foreigner, and another hinted that I was mad. But I put up with every rebuff, and went on with my plans, *determined not to be put down.*" Mr. Nicholas Wood, who was present during Stephenson's examination as to the possible speed of locomotives, afterwards gave it as his opinion that if twelve or fifteen miles had been stated as practicable, not a single person present would have believed the announcement!

The opponents of the Liverpool and Manchester Railway Bill had a mute but eloquent advocate in their behalf, and they did not fail to use it to advantage. Chat Moss, a peat bog extending to about twelve square miles, and in the direct line of the railway, offered an argument of the most convincing character against the scheme. Mr. Harrison, one of the opposing barristers, thus accurately described it:—"Chat Moss rises in height, from the rain swelling it like a sponge, and sinks again in dry weather; and if a boring instrument is put into it, it sinks immediately by its own weight. The making of an embankment out of this pulpy wet moss is no very easy task. Who but Mr. Stephenson would have thought of entering into Chat Moss, carrying it out almost like wet dung? It is ignorance almost inconceivable. It is perfect madness in a person called upon to speak on a scientific subject to propose such a plan.

The idea of working the traffic over the proposed railway by means of locomotives was thus ridiculed by the same learned gentleman:—"When we set out with the original prospectus, we were to gallop, I know not at what rate; I believe it was at the rate of twelve miles an hour. My learned friend, Mr. Adam, contemplated—possibly alluding to Ireland—that some of the Irish members would arrive in the waggons to a division. My learned friend says that they would go at the rate of *twelve miles an hour*, with the aid of the devil in the form of a locomotive, sitting as postilion on the fore horse, and an honourable member sitting behind him to stir up the fire, and keep it at full speed." The learned counsel then attempted to prove that it was *impossible to attain a speed of six miles an hour!* The following was one of the reasons which he gave in support of his theory. He said:—"Locomotive engines are liable to be operated upon by the weather. You are told they are affected by rain, and an attempt has been made to cover them; but the wind will affect them; and any gale of wind which would affect the traffic on the Mersey would render it impossible to set off a locomotive engine, either by poking of the fire, or keeping up the pressure of the steam till the boiler was ready to burst."

The testimony of civil engineers of eminence fully bore out the taunting criticism of barristers. "No engineer in his senses," said one of these gentlemen, "would go through Chat Moss if he wanted to make a railway from Liverpool to Manchester." With such an opinion expressed on professional authority, Mr. Alderson was justified in employing all the resources of his legal ability and great eloquence in a two days' speech against the Bill. He declared the scheme to be the most absurd that ever entered into the head of man to conceive. A majority of the committee believed the statement of the learned counsel; and the measure was withdrawn after a bitter and determined contest extending over two months.

The prospect held out to Stephenson after this crushing defeat was indeed a dark one. At various times during the proceedings he had been dubbed a madman and a fool, and even his friends had wavered in their faithfulness towards him. His disappointment and chagrin were intensified when he found that the projectors of the railway had resolved to renew their efforts in another Parliamentary Session, but had expressed their intention to call to their aid engineers only of the most distinguished status and repute. That was done by the appointment of Messrs. George and John Rennie to the position which Stephenson had lately filled, and these gentlemen proceeded to carry out another survey.

But the Killingworth engineer, the locomotive and the

railway, were soon to emerge from the cloud which enveloped them. On the completion of the new survey, the plans were prepared, and a second Bill was presented to Parliament, which passed without any undue protraction of the arguments or examination of witnesses. This result was effected mainly through the measures which wére taken by the promoters to conciliate the favour of their most formidable opponents. By so altering the route as to render inviolate the precincts of certain estates, the antagonism of the owners was withdrawn; while a questionably generous concession was made to popular prejudice against the locomotive, by *offering* to eliminate from the Bill any clause which would empower the company to use travelling engines upon the line.

Having obtained powers to construct the railway, the appointment of the chief engineer next engaged the consideration of the directors, and they decided to continue the services of Messrs. Rennie, in order that these gentlemen might superintend the operations, while to Stephenson would be given the post of resident engineer—a subordinate position, but one, under the circumstances, which demanded in its holder the possession of considerable ability and practical experience. But professional etiquette interposed a barrier to that arrangement; Mr. George Rennie stipulating for the appointment to the inferior office to be left in his own hands. To that stipulation the directors could not agree, and so the principal engineership was offered to Stephenson, with its emolument of £1000 a-year. The offer was accepted, and he removed to Liverpool, where he at once made arrangements for carrying out his stupendous undertaking.

CHAPTER IX.

THE VICTORY OF THE "ROCKET."

WHEN Stephenson entered upon his duties as engineer-in-chief of the Liverpool and Manchester Railway, he had no trained staff of assistants or draughtsmen. As resident engineers, he selected John Dixon, whose connection with the Stockton and Darlington line has already been noticed, Joseph Locke, and William Allcard. To these gentlemen were allotted the three sections into which the railroad was divided—the former having charge of the most difficult portion, namely that over Chat Moss. After taking an active part in the organisation of the Forth Street Engine Works, Robert Stephenson had sailed for South America, in June 1824, having being engaged by the Columbian Mining Association to direct the operations of that company at Mariquita. George Stephenson was thus left to face the difficulties of his new position without that aid which his promising son *would undoubtedly* have been able to render had he re*mained in England.* The principal draughtsman, Thomas

Gooch, was one of a number of pupils who were employed in office-work, and in learning their profession under the first railway engineer. Gooch also acted as Stephenson's secretary, and was able to bear personal testimony to the unflinching zeal and perseverance of his chief. "Like master like man," was an adage which was amply illustrated in the careers of Stephenson's pupils, for they all, without exception, attained distinction in their profession, and severally exemplified the teaching of their instructor, whose frequent advice to the young men was:—"Learn for yourselves; think for yourselves; make yourselves masters of principles; persevere, be industrious: and there is then no fear of you."

Chat Moss, that huge peat bog, tried to the utmost the patience, energy, or skill of all who were concerned in carrying a railway across its pulpy, spongy surface. For ages, year by year, mosses had grown and withered—the partially decayed stratum of one season forming a bed for the growth of the following, until the whole made up a submerged mass of soft slimy peat, with a depth of from twenty to thirty feet, and extending over an area of about twelve square miles. While some engineers of eminence had pronounced the construction of a railroad over such a foundation to be wholly impracticable, others had fixed the cost of the line at a figure which would have destroyed any prospect of interest being earned in return for the invested capital, had the estimates proved at all accurate. Undaunted by the unfavourable opinions of others regarding his plan, or the formidable nature of the task itself, Stephenson commenced the operations. Throughout the progress of the work his whole bearing and conduct formed a practical commentary upon his own homely precepts:—Learn for yourselves; think for yourselves; master right principles; persevere, and be industrious!

The theory conceived by the engineer, and upon which he designed to carry the line over Chat Moss, was that of a *floating roadway* of sufficient strength and buoyancy to carry the rails and their moving weights securely. The digging of drains along the sides of the proposed line was first attempted, but soon had to be abandoned in many places on account of the trenches quickly filling up with the slushy material through which a cutting had been made. In the endeavour to form a pathway for the workmen and boys engaged, the engineer was more successful. A thick platform of broom or heather, laid upon the surface of the Moss, constituted the foundation upon which were laid the temporary sleepers necessary for the rails. Along the railroad thus formed the materials required for making the permanent way were carried in waggons propelled by boys. The principle of the permanent road was but an extension of that of the temporary waggon-way; the foundation of the line on the softest portions of the Moss being constructed by intertwining heather between the bars of strong wooden hurdles, which were laid down as a bed for the sleepers, drains, and rails. To prevent any undue sinking of the railroad at these slushy parts, the spaces between the sleepers were packed with dried peat or heather, instead of ballast or gravel, as is usual in the construction of ordinary railways.

Never at a loss for an expedient when beset by a difficulty, Stephenson caused tar barrels to be laid down on either side of the Chat Moss railroad, at parts where the wet nature of the bog interfered with the cutting of drains.

The barrels were placed end to end, and properly secured, so as to form wooden sewers for the drainage of the line. This device had the desired effect, but a more formidable difficulty presented itself when a junction between the floating road and *terra firma* came to be wanted. For a distance of four miles the railway was upheld by its own buoyancy, like a long strip of cork on the wet and spongy heath. Now the ends of the strip must be firmly joined to the solid ground on the extreme edges of the Moss. At the edge nearest to Liverpool that was effected with but little difficulty. The Manchester end of the floating line was not, however, so easily secured. Thousands upon thousands of cubic yards of earth and dried peat were thrown into the insatiable bog, but without apparent effect. This went on for weeks, but no trace of an embankment appeared to reward the directors and engineer for their outlay and labour. The projectors naturally grew impatient at what seemed to be a fruitless expenditure of capital. Those who had sneered at the courageous idea, and ridiculed the engineer with whom it had originated, were now jubilant at the prospect held out of a veritable fulfilment of their prognostications. Stephenson was taken to task one day by an over-zealous director of the company, who instanced the energy of the great Napoleon as worthy of imitation by the engineer in such a crisis as he then found himself. Stephenson's reply was characteristic of the man: "Never mind Napoleon! With money, materials, and men, I can make a railroad over Chat Moss. We must keep filling in. There is nothing else for it. The stuff is doing its work out of sight, and it will soon show, *if you will only have patience!*"

Many railways of later date than the one between Liverpool and Manchester afford examples of the successful accomplishment of greater engineering tasks. It is true that the scheme of a line over Chat Moss was declared utopian by many who were qualified to express an authoritative opinion. The Sankey Viaduct, with its nine arches, each of fifty feet span; the Olive Mount Cutting, the depth of which is in some parts eighty feet, while its rocky length extends to nearly two miles; the Liverpool Tunnel, hewn and blasted through solid rock, or cut through beds of yielding sand, for a distance of two thousand two hundred yards: these undertakings all demanded great ability in their designer and constructor. All have been repeatedly eclipsed, however, by more modern achievements in engineering. But it should be borne in mind that when the Killingworth engine-wright entered upon the practical portion of his duties at the time now under review, he had not only to devise and provide the accessories and implements necessary to a railway contractor, but he was also called upon to organise and drill a band of navvies specially for the work of railway making. For that end the practice he had acquired as a ballast-heaver at Willington Quay peculiarly fitted him. He had not forgotten the art of handling a shovel effectively. The railroad navvy can trace his history no further back than the time of George Stephenson, who first taught him the use of the pick, the shovel, and the barrow.

Besides the smaller details of stations, bridges, girders, crossings, and sidings, and larger works connected with the permanent way, which were for the most part planned under his immediate direction, Stephenson had to design *fixed engines* and rolling-stock for the working of the railway. It was not until the line was considerably advanced that the directors seriously debated the question of using locomotives. A deep-seated prejudice existed in the minds of many regarding the travelling engine. The Stockton and Darlington Company had experienced some difficulty with proprietors whose estates joined that line, in regard to the excessive quantity of cinders which "Locomotion" and its companion "Hope" emitted from their funnels. Law proceedings had resulted from the alleged nuisance. The minority of the directorate who opposed the introduction of the locomotive upon the second public railway had some show of reason, therefore, for the opposition in the bad repute of the engines upon the South Durham line. But amid the harassing work of his every-day life, Stephenson found time and energy to fight the battle of his favourite motive power, and he proved successful in the encounter.

At the urgent request of his father, Robert Stephenson returned to England in 1827, and at once assumed the direction of the affairs of the Engine Works at Newcastle-upon-Tyne, which had been allowed to get into confusion during his absence. At one time the prospect of success for the manufactory appeared so hopeless that the stoppage of its operations had been seriously entertained. But now, with his son reinstated in his old position as manager, the engineer of the Liverpool and Manchester Railway felt that could he but succeed in getting his directors to try the locomotive for themselves, and upon their own line, its ultimate triumph was assured. In order to attain the consummation of his great desire, he never let slip an opportunity for urging the matter upon the attention of those who were most disinclined to give their consent to his project. In season *and out* of season he pressed forward the claims of *his foster-child,* until at length, wearied by incessant importunity, the Board ordered a travelling engine to be built, in order, as Stephenson alleged, that it might transport expeditiously from one point to another, as might be required, the various materials used in the construction of the line. That locomotive was forwarded from Newcastle in 1829, and the work performed by it justified the advice which had been given by the engineer.

With an engine travelling upon their own railway now rapidly nearing its completion, one might have imagined that the question of its utility, or the reverse, could have been easily determined by the directors. Not at all. The inventive genii of Great Britain and other countries seemed to have conspired to harass and bewilder the unfortunate gentlemen, who were deluged with all sorts of schemes for the propulsion of their future trains. Worn out by the continual applications, correspondence, interviews, advice, and journeys of inspection which they had to undergo, as well as by the persistency of their chief engineer, the directors finally succumbed to the inexorable decree of fate, and a prize of £500 was offered for a locomotive which should be proved upon trial to have fulfilled the following conditions, which were publicly circulated:—

"The engine must effectually consume its own smoke. If of six tons weight, it must be able to draw, day by day, twenty tons weight (including the tender and water tank) at ten miles an hour, with a pressure of steam on the boiler not exceeding fifty pounds to the square inch. The boiler must have two safety-valves, neither of which must be fastened down, and one of them be completely out of the control of the engineman. The engine and boiler must be supported on springs, and rest on six wheels; the height of the whole not exceeding fifteen feet to the top of the chimney. The engine, with water, must not weigh more than six tons; but one of less weight would be preferred, on its drawing a proportionate load behind it; if only four-and-a-half tons, then it might be put on only four wheels. The company to be at liberty to test the boiler, etc., by a pressure of one hundred and fifty pounds to the square inch. A mercurial gauge must be affixed to the machine, showing the steam pressure above forty-five pounds per square inch. The engine must be delivered, complete, and ready for trial, at the Liverpool end of the railway, not later than the 1st of October 1829. The price of the engine must not exceed £550."

Although many attempts had been made by several hands, to fulfil the requirements announced by the directors, only four steam locomotives were present on the ground allotted for the trial when the day appointed dawned. Various causes conduced to that result. In some instances engines had been constructed, which, upon examination, were found to be so defective in principle as to render the chances of their success very small indeed, and the machines were consequently not entered for competition. In other cases, repeated alterations had delayed the finishing of engines, until the time limited for entry had expired. The four exhibited were the "Novelty," "Sanspareil," "Rocket," and "Perseverance," built respectively by Messrs. Braithwaite & Ericsson, Timothy Hackworth, R. Stephenson & Co., and Burstall.

The memorable trial of locomotives at Rainhill began on the 6th of October 1829, and extended to the 14th, on which day the prize was awarded. Although not the first on the list for competition, Stephenson's "Rocket" was the earliest to be ready for starting, and it was accordingly ordered by the judges to proceed on a preliminary trip. Neither the gentlemen more nearly concerned in the trial, nor the spectators in general who witnessed it, were at all prepossesed in favour of the engine from Newcastle. Its appearance was against it. It looked as if it were all funnel—a stunted body with a long, very long neck.

Along a level stretch of railroad, two miles long, each engine was required to make twenty double journeys during the day, at an average speed of not less than ten miles per hour. The first trip of the "Rocket" was quite satisfactory, as it accomplished the distance of twelve miles in about fifty-three minutes. Still, as people said, its appearance was against it.

The second engine ordered out by the judges was of a totally different character. It looked compact and handy, and its lines were harmonious and in keeping with the purpose for which it had been built. The "Novelty" was not only a favourite with the dense crowd of spectators, but the majority of the judges were also disposed to regard it with preference. Its water and fuel were carried without the aid of a separate tender, and the weight of the whole was but a little over three tons. While travelling its experimental journey, it occasionally moved at the rate, per hour, of twenty-four miles.

Timothy Hackworth, the designer of the "Sanspareil," which next passed in review before the judges at Rainhill, laboured under great disadvantages in the construction of his locomotive. He was a man far in advance of his time; and George Stephenson, upon more than one occasion, was indebted to his friend for sound practical advice when beset by engineering difficulties. Trained under William Hedley at Wylam, Hackworth had acquired strong sympathies in favour of the travelling engine; and when the competition was advertised, he at once took steps to enter an appearance with an engine that would not disgrace the future reputation of its contriver. But he had not the facilities for carrying out his idea without the assistance of others. He was at that time resident engineer and manager of the

Stockton and Darlington Railway, and his duties in that capacity prevented him from giving that personal oversight to the building of the "Sanspareil," the want of which, in all probability, told greatly against that engine's efficiency at the trial. The cylinders were made by R. Stephenson & Co.; tho boiler at the Bedlington Iron Works.

The second day of the contest was marked by the failure of the blast-bellows of the "Novelty"—a contrivance that was unique in its way, but one that constituted a weakness in the neatly-built and favourite engine—and it could not in consequence be exhibited. The boiler of the "Sanspareil" also showing a defect, Hackworth's locomotive had to be withdrawn for a time in order to be repaired. As for the "Perseverance," it was unable to travel faster than six miles an hour, and had therefore failed in the important condition regarding speed.

One of the four competitors having thus been definitely disposed of, and two undergoing repairs, considerable dissatisfaction began to be manifested by the crowd which had again assembled to witness the singular spectacle of a locomotive race. Stephenson noticed the gathering storm, and immediately endeavoured to allay popular feeling by running out the "Rocket," to which was attached a passenger coach. Into the latter thirty persons were huddled without much ceremony; and the engineer himself acting as driver, the engine at once started along the line at the rate of nearly thirty miles an hour. That performance astonished alike the travellers, the spectators, and the judges. It was not, however, until the following day, the 8th of October, that Stephenson's bald and unpopular locomotive made any decided progress in public favour.

In its final trial the "Rocket" more than sustained

character which had been expected from it by its designer. With a load of thirteen tons weight in waggons, travelling alternatively forwards and backwards along the two-mile course, the engine accomplished a distance of thirty-five miles, with the necessary stoppages for reversing the motive gearing, in one hour and forty-eight minutes. A second journey of the same length was performed in fourteen minutes longer than the first series of trips. The average velocity attained over the whole seventy miles was fifteen per hour; the maximum speed for the same time being twenty-nine. Ere the task of the "Rocket" had been concluded on the third day of the competition, people discovered excellencies in the hitherto despised travelling engine that were not before apparent to the popular eye. In short, as some one said, "*She did not look so bad after all!*"

The "Novelty" having been repaired, was again tried upon the 10th of October, when its appearance and style of working as it started upon the first trip for that day, gained for it a renewal of some portion at least of the partiality which it had lost through its mishap. But the sudden bursting of one of its pipes ended the hopes of its owners and their chance of gaining the offered prize. Three days afterwards, the "Sanspareil" was similarly unfortunate; for during the eighth trip, made at the average rate of fourteen miles per hour, the engine broke down through a failure in its water-pump, and the prospects of success for Hackworth's engine were also at once extinguished.

The "Rocket" being the only competing locomotive which had fulfilled the stipulations in terms of which the trial was conducted, R. Stephenson & Co. were declared to be the [illegible] *the prize of £500.* That the public concurred in the verdict of the judges, is apparent from the shares of the company immediately rising in value to the extent of ten per cent. The "Sanspareil," although it did not gain for its designer much credit during the memorable trial, was nevertheless a good and serviceable engine, as will be seen in the fact that it was employed for many years upon the Bolton and Leigh Railway. The following notice of the famous "Rocket" and its veteran progenitor, "Puffing Billy," of Wylam celebrity, appeared in the *Illustrated London News* of the 15th of October 1864:—

"On the south side of the Museum are a couple of locomotive engines—old, very old—worn, bent, bruised, and rusty, but full of interest; for one is the oldest locomotive engine in existence, the parent of all that have since been produced—the original old 'Puffing Billy' of the Wylam Colliery, constructed there, in 1813, by William Hedley for Christopher Blackett, whose name will ever be remembered as one of the earliest and staunchest friends of the locomotive. This 'Puffing Billy' was not the first engine of Mr. Blackett. He had tried Trevithick's and other inventions, and had been nearly blown up several times, but at last this homely-looking, sturdy machine was perfected, and from that date—1813—up to a comparatively recent period, it has been dragging—very slowly, it is true—heavy loads of coals and waggons to and fro. It has a battered, and bruised, and hard-worn look, with indubitable marks of extreme old age about it; but it was a good, hard-working, willing drudge, and has well-earned the quiet time it will have in future here in the Kensington Museum. Upon one side of it hangs a document, dated 1815, from which it appears some great objection had been made by local landowners to the puffing propensities of the aforesaid 'Billy,' and a case is referred for opinion to a barrister, named Williams, whose opinion is that there is no objection, arising from the lease itself, to Mr. Blackett conveying his coal-waggons by means of this steam-engine; but he thinks that the use of such an engine may be deemed a nuisance if the smoke or noise of the engine disturbs the cattle grazing on the lands adjacent to the *waggon-way.* The locomotive had a narrow escape of being put down *as a common* nuisance. We might now as easily think of putting down earthquakes or eruptions of volcanoes as putting down the locomotive engine; but in those days it was different, and had an extinguisher been put on 'Puffing Billy' at that time, the introduction of locomotives and railways for passenger traffic might have been delayed for many years. Beside this venerable specimen of mechanical ingenuity stands another locomotive engine, worn, bruised, strained, and cracked like its neighbour, but not so old as 'Puffing Billy,' which dates as far back as 1813, while this did not come into existence until the year 1829. These engines are well placed side by side, for the spectator may, by looking at them in such a position, easily come to several important conclusions. In designing the old engine, it is evident that its inventor never dreamed of any speed worth thinking of. To crawl along, with a heavy load behind it, was all that this engine was expected to do, and it did this patiently, but with a great deal of fuss and clatter, for half-a-century; and for such a performance, and considering its extreme old age, it is entitled to our highest respect. But, turning to its neighbour, we see that its inventor and constructor did think of high speed; there is a wonderful look of speed about it, and it did actually travel at forty miles per hour, for it is the celebrated 'Rocket' engine of George Stephenson, the parent engine of our present locomotive system. This is really a very interesting relic, for it was the great starting-point of high speed and passenger railways."

The victory of the "Rocket" effectually silenced the clamour of the more bitter opponents of the railway system. It also settled the vexed question of the application of steam locomotion for passenger traffic. True, many could not overcome their prejudice against the new mode of travelling; but such persons soon found themselves in a most unenviable minority. The tables had been turned, and the opponents, not the advocates, of railroads were regarded as fit and proper candidates for admission into asylums for the insane!

CHAPTER X.

THE TRIUMPH OF THE RAILWAY.

WHILE the Stockton and Darlington line was in course of construction, its engineer uttered a prophecy regarding the future of the railway system. Addressing his son Robert, and his assistant John Dixon, George Stephenson said :—"Now, lads, I venture to tell you that I think you will live to see the day when railways will supersede almost all other methods of conveyance in this country: when mail-coaches will go by railway, and railroads will become the great highway for the King and all his subjects. The time is coming when it will be cheaper for a working-man to travel upon a railway than to walk on foot. I know there are great and almost insurmountable difficulties to be encountered, but what I have said will come to pass, as sure as you live. I only wish I may live to see the day, though that I can scarcely hope for, as I know how slow all human progress is, and with what difficulty I have been able to get the locomotive thus far adopted notwithstanding my more than ten years' successful experiments at Killingworth." Much of that prophecy has already been fully accomplished. The portion which relates to the cheap conveyance of the working classes has only been partially fulfilled!

The completion of the Liverpool and Manchester Railway was considered an important national event. By the 1st of January, 1830, a single line of rails had been laid down throughout the whole distance; and on the 15th September of that year the opening ceremony took place, the railway being then finished and ready for traffic. The Duke of Wellington, who was Prime Minister, honoured the occasion with his presence; as did also the Home Secretary, Mr. Peel, and an immense number of distinguished personages. Many scientific gentlemen of this and other countries were likewise present to satisfy themselves by ocular demonstration, that what had been considered to be impossible had actually been performed, through the indomitable energy, industry, and perseverance of a humble, self-taught enginewright, from a small, out of the way colliery village in Northumberland!

A sad calamity marked the inauguration of the line. Eight locomotives, which had been built at the Forth Street Works, formed, with their several trains, an imposing procession that was cheered vociferously by the thousands who watched its progress. On arriving at Parkside, seventeen miles from the starting-point at Liverpool, the leading engine, called the "Northumbrian," and attached to a carriage which contained the Prime Minister, drew up on one line, so that the trains which followed might pass along the other in review before his Grace. Mr. Huskisson, one of the members of Parliament for Liverpool, and who had taken much interest in the projection of the railway, suddenly crossed the rails for the purpose of speaking to the Duke, when the "Rocket" was seen approaching rapidly. The warning cries of the bystanders only served to confuse the honourable gentleman, and before he could recover his presence of mind, and step clear of the line, the engine was upon him, inflicting injuries from which he died a few hours after the accident. Notwithstanding the gloom cast by that melancholy circumstance, the triumph of the railway system had been assured.

A large passenger traffic immediately sprang up to reward the directors for their outlay, and to demonstrate the soundness of George Stephenson's conclusions. The stage-coaches running between Liverpool and Manchester had carried on an average about four hundred and fifty persons per day. It was calculated that half that number might be induced to transfer their custom to the railway, but the line had only been opened a very short time when the daily number of passengers was found to exceed twelve hundred! Even in the first year of its existence, the railway which had been prophecied to be fraught with devastation to property and ruin to shareholders, earned an annual profit of eight thousand pounds; while the adjoining land rose so much in value, by reason of such proximity, that when the company required to purchase some ground for the extension of stations and sidings, it was found that the price had increased considerably.

The Liverpool and Manchester Railway, in its projection and construction, had been the field upon which the battle for and against the locomotive had been fought. The visitors followed up their success by engrafting upon each new engine placed upon the road the latest improvements which their growing experience suggested. It had been incontestibly shown that a railway could be made over ground presenting great natural obstructions to the work of the engineer. It might have remained for financial results slowly to determine whether the extension of the system should be proceeded with, or otherwise left as an open question for the future to decide. But speculators were encouraged by the brilliant prospect that was at once disclosed by the completion of the Liverpool and Manchester line to embark without much delay in fresh undertakings. Railroads were speedily projected between the great centres of industry and wealth; while the profession of a railway engineer offered dazzling inducements to many, whose qualifications were fanciful or real, to enter its ranks. The name of George Stephenson, however, had become as a household word in connection with the new medium of transit, and his services were not only in demand for the great lines which were proposed in his own country, but continental railway companies sprang up also to claim an interest in his genius and ability.

His sterling honesty and sound common sense were conspicuous during the period of the railway mania which infested English society in the years 1845–46. Intoxicated with the desire to get rich, or richer at a bound, men lost their mental equilibrium, and plunged into the wildest of schemes. To the honour of Stephenson's memory be it said that no shareholder could ever trace his own ruin, or the beggary of his family, to the collapse of a specious project to which the great engineer had lent his name. "Will such a line pay?" was the invariably first question which he endeavoured truthfully to answer. Engineering difficulties were accounted by him as but of secondary moment. Whenever he could not satisfy his mind that any new venture offered a reasonable expectation of success financially, he at once, regardless of selfish interest or of consequences to others, gave it his uncompromising opposition. Thus we find him holding aloof from all the reckless schemes that were put forward by too sanguine or unprincipled promoters. Although he was frequently offered considerable sums, if only he would allow his name to appear in the prospectus of a dubious project, he stedfastly refused to truckle to dishonesty. George Stephenson would always render solid service for adequate payment. He never touched a questionable bribe.

The harassments and rebuffs which he had endured while fighting the battle of the locomotive and the rail had impaired the natural kindness and sympathy of his disposition to some extent, and rendered him at times, previous to the hour of his great triumph, particularly sensitive of the slightest opposition to his wishes: irritable, impatient, exacting. But after the opening of the second public railway—when the feverish fervour of open foes had been transformed into the fawning flattery of so-called friends—his former amiability and genial temper returned, and continued to be conspicuous traits in him until the closing scene in his eventful life. Not that he was susceptible to the adulation of such as are ever ready to run to the assistance of a man when he requires no help. Stephenson could estimate such a favour at its true value. But having attained the great object for which he had struggled and toiled, almost unaided and alone, he could now look with comparative complacency, if not indifference, upon all the troubles and trials incident to his position.

Associated more or less with his son Robert, Stephenson continued to devote his surpassing powers of mind and

Spotted something that isn't quite right?

Let us know and we'll fix it for future reprints.
And as a thank you for your diligence, we'll
send you a free Sothis book of your choice.

publisher@sothis.press

This page intentionally left blank

Except of course for that line above. And this one.
And whilst we have your attention, perhaps you'd
like to follow us on Instagram @sothis.press. Go on.

Made in United States
North Haven, CT
06 November 2026

21878604R00076